Family Fantasies and Community Space

Family Fantasies and Community Space

STUART C. AITKEN

RUTGERS UNIVERSITY PRESS
New Brunswick, New Jersey, and London

Library of Congress Cataloging-in-Publication Data

Aitken, Stuart C.
    Family fantasies and community space / Stuart C. Aitken.
       p.   cm.
    Includes bibliographical references and index.
    ISBN 0-8135-2461-X (alk. paper). — ISBN 0-8135-2462-8 (pbk. :
alk. paper)
       1. Family—California—San Diego—case studies. 2. Partiarchy—
California—San Diego—Case studies. 3. Community life—
California—San Diego—Case studies. 4. Human geography—
California—San Diego—Case studies. 5. Spatial behavior—
California—San Diego—Case studies. I. Title
HQ536.15.C2A57   1998
306.85'09794'985—dc21   97-17634
                                                          CIP

British Cataloging-in-Publication information available

Manufactured in the United States of America

To Peg, my partner in all of this

Contents

Acknowledgments

Research for this book was supported in part by Grant SES-9113062 from the National Science Foundation and a grant from the San Diego State University Foundation. Special thanks go to the students who were employed on the study: Marta Miranda, Chris Carter, William Granger, Leslie Bolick, Suzanne Michel, Thomas Herman, Katina Pappas, Nickolas Deluca, Meg Streiff, Susan Mains, Matt Carroll, Serena McCart, and Pauline Longmire. We all agreed that the interviewing was, for the most part, fun and the discussions stimulating, but we would prefer to leave the transcription of interview tapes to someone else.

I would also like to acknowledge the support of faculty and student seminars and reading groups, particularly the interdisciplinary discussion groups that were supported by the College of Arts and Letters and the Department of Geography at San Diego State University. For me, these groups provided forums for working through complex perspectives and ideas on culture, politics, gender, and sexuality, some of which eventually found their way into this book.

In addition, I would like to thank all the families in San Diego who filled out our questionnaires or agreed to be interviewed (or both) as part of the study that drives large parts of the arguments in this book. Any opinions, findings, and conclusions or recommendations expressed in this book are mine and do not necessarily reflect the views of the National Science Foundation, San Diego State University, or the students and families involved in this project.

Participant Anonymity

Fictitious names are used to ensure the anonymity of the nine families whose interview materials are used extensively in this book. The names of any individuals whom they mention in interview excerpts are also fictitious. Excerpts from interviews of other participants in our study are identified simply by the status of the family members who are being quoted (a mother, a father, an uncle, and so forth). The names of small neighborhoods, communities, and institutions have been changed if there was a possibility that the participants' real identities could be uncovered.

Family Fantasies and
Community Space

Encountering
Family Fantasies

My Part in All of This

Somewhere it all began to fall apart. I do not know exactly why I became uneasy with my feelings about family life and my role as a father and husband, but it was sometime between the birth of Ross in 1990 and Catherine in 1993, and it was after we found a home in which to raise our growing family. This vague time and indistinct space constitute a fuzzy destabilization of my life world. Before this period and place in my life, I thought I knew what constituted "family," and I imagined the ways in which I would like to see myself and my family nested within community life. With the recent attainment of a tenured position at the university, my job as a provider, it seemed, was gaining significant credibility. Peg, my partner in all of this, was surprisingly comfortable with the idea of motherhood, although I was somewhat uneasy that she wanted to put her career on hold for a few years of "raising children." My unease did not stem from financial strain, although Southern California is certainly not conducive to hedonistic family living on the income of an associate professor at a state university. I was much more concerned about the compromise of my feminist and antipatriarchal ideals. I would like to think that my feminism is more than what Liz Bondi (1990) calls "gender tourism," wherein white male academics contrive to "get a bit of the other" (Moore 1988) by colonizing and appropriating the exotic discourses of women and minorities while still maintaining an escape route back to the "fatherland" if things get too uncomfortable. I am genuinely concerned about how we construct our identities, how they are linked to notions of family, community, and society, and how we use these identities to make sense of people, places, and events. My opposition to a system rooted in white male authority and female subordination does not arise from a vague need to be politically correct but, rather, from a deep-seated anxiety about my own sexual and social being. Men are, after all, also gendered within a phallocentric system, and they too are subjugated by "the law of the father."

My unease with my identity as a father and a husband grew as I perceived contradictions between my family and my work. How could I teach and write about the myth of the nuclear family when I was living that ideal? I was reading a considerable amount of postmodern and psychoanalytic literature as well as immersing myself in feminist and critical social theory. I knew about some of the "evils of patriarchy," like the exploitation of mothers as domestic slaves and their spatial entrapment in suburbia. I believed that the notion of the nuclear family harbored dangerous gender power structures. I was determined to avoid traditional male roles. I wanted not only to be present for my family but also to be a nurturing and caring role model who was patently, if not parent-ly, domesticated. Candidly, I charted the pitfalls of my parents and was ready to give them a wide berth, but this exercise left me with no concrete role models. In a vain attempt to loosen the fetters of my sex, I explored the mythopoetic images of 1990s fatherhood and maleness. I was moved by some of the poetry of Robert Bly, but most of the mythopoetic men's literature left me with a feeling of unease because it either constructed a largely womanless past or fell into the Freudian trap of blaming the mother.

We chose to live in a community that bore little resemblance to suburbia: a multicultural, socially and economically diverse residential area close to the university. Peg was not entirely comfortable with the choice, but we both felt pressured by the prospect of ever-increasing housing prices and unpredictable interest rates. With a tight budget we could afford this house in this place at that time. And so, within our new community, I began to map out a family fantasy that seemed to encompass our way of life.

Again, Peg began to feel uncomfortable. She became nervous about

my hanging around the house at odd hours during the day. She was suspicious of my attempts to "help out" with the kids on a day-to-day basis. She confronted me with the absurdity of merely helping out in our
partnership rather than taking responsibility for stressful tasks such as
organizing day care, providing education, and meeting medical needs.
So I also tried to become involved in the long-term responsibilities. As
I freed up time at the university to be with the kids, Peg became concerned about my commitment to work and my responsibilities there. I
was astounded: here I was agonizing over what I needed to do at home,
and she worried about what I was not doing at work! How dare she
involve herself with my life at the university, which I felt was going
well? Numerous articles were in print, my first two books were "in press,"
and I was close to securing promotion to full professor.

Nonetheless, our home life was deteriorating. Peg and I had increasing difficulty communicating our needs and expectations around gender roles and relations. In those many moments of fatigue, frustration,
and insanity, when there was no time to rationalize my actions and
thoughts, I found myself disciplining my children and shaming my wife
in ways that I had sworn I would never do. I began to detach myself
emotionally from the family. Half-remembered images of my father's
emotional absence from my upbringing caused me great pain. The re-
alization was glacially slow, but by the time Ross was four years old, I
was aware that I did not know what I was doing as a father. Peg felt
abused and subjugated, her body and her mind chained to the ideology
of familism. I felt betrayed by my feminist ideals and liberal, postmodern
views on life. I knew that abstract academic concepts did not constitute day-to-day living. I was reading enough postmodern critique to be
uncomfortable with metanarratives, whether they were Marxist, feminist, or New Age. I knew the dangers of thinking that I stood on a stable
hilltop, overlooking family life in a detached Archimedean way. Peg
and I were in the thick of the emotions and experience of raising children and trying to keep a marriage together. I sought solace and guidance from critical social theories and found an unsettling image of
masculinity embedded within the popular, New Age "male"-order cata-
logues. Clearly feminism and postmodernism provide a valuable critique
of life under late capitalism by de-centering logic and reason and replacing them with emotion and experience, but they do not (and should

not) offer utopian ideals of a "brave new world." Critical perspectives legitimately require us to construct our perceptions of the world anew, in ways that question the obvious and "natural," but they cannot offer answers without undermining a central set of premises that are suspect of solutions. Why does society still uphold the "naturalness" of a two-parent family under one roof or family values that are entrenched in the constraining myth of the nuclear family? Despite academic and in-stitutional lip service to other family fantasies, it is still the powerfully constituted norm of the family writ large that brings the four-year-old son of a single parent home with the question "What is wrong with our family?" What is so unnatural about a single-parent family? More to the point, is there anything "natural" about raising children?

In the middle of all the personal trauma, responsibilities, miscom-munication, conflict, love, anger, joy, and hurt I received a grant from the National Science Foundation (NSF) to study the geography of fami-lies with young children. My knowledge of cities and communities, my supposed expertise in interviewing methods, and my understanding of spatial and feminist theory constituted the cornerstone of the NSF pro-posal, in which I placed myself as an impassioned and somewhat aloof observer of other families. I did not realize until later that I was really positioning myself as a voyeur. I became involved in the lives of the families participating in our investigation because I needed to know how to make my family work. Armed with graduate student help, I followed the lives of several hundred people over a three-year period. We got in touch with women who were pregnant and asked whether they and the other adult members of their households would fill out our question-naires. If they were expecting a first child, we asked for in-depth inter-views. Today I am still in touch with many of our participants. I am no closer now to finding out how things work than I was when we began. I know now that there are as many problems and solutions as there are parents and children. Many contradictions in actions and beliefs remain unresolved for me, but it may not be appropriate that they find resolu-tions. People should live their lives without being fettered by someone else's elevated and overbearing concept of what constitutes a family.

New Images of Families and Communities

The accounts in this book are of men, women, and children who are living out complex and subtle family geographies, but one common ingredient of change for these families was the birth of a child. The form of family here is limited. No gay or lesbian couples, for example, surfaced in the NSF study. My students and I contacted women who were pregnant and had been for a first appointment at one of eight obstetrical-gynecological clinics dispersed throughout the San Diego metropolitan region (see appendix A). Households that agreed to participate in the study had at least one member employed at the time of our first survey, although some participants lost their jobs and medical insurance during the course of the study. The sample as a whole does not have a representative number of extremely low-income or extremely high-income families. Because the "gatekeeper" who provided access to the family was always the mother, there are no single fathers in our study.

My primary intent was to procure a sample that was geographically representative of communities in San Diego, and this goal was, to a large extent, achieved.[1] The study comprised a questionnaire sent to geographically dispersed households within which the woman was expecting a child and two follow-up surveys sent prior to the child's first and third birthdays. A total of 577 adult members of the households contacted agreed to complete our questionnaire and participate in the study for three years; 231 completed the second survey, and 166 remained for the full duration of the study. All three surveys focus on household members' day-to-day activities, responsibilities, and opinions about family life. The follow-up surveys assess changes in gender roles and relations after the birth of the child. Within this sample population, we conducted 127 in-depth interviews of adults in households expecting a first child. These face-to-face interviews provide qualitative information on specific day-to-day events, household members' feelings, attitudes, and interdependencies.[2]

The qualitative materials provide the primary backdrop for the theoretical arguments I make throughout the book. The interview materials are important to the theoretical threads that run through the book not because of their scientific merit but rather because of the glimpses, however fleeting, of change and resistance in contemporary family life that they offer.

CONVERSATIONS AROUND THE BIRTH OF A CHILD

The birth of a child, more than many other kinds of events, stimulates the kind of changes that might highlight the "structural fragility" of family life (Stacey 1990). Individuals seek to make sense of events in their lives and of their surroundings and to define and locate themselves with respect to those events and surroundings. An extraordinary event such as the birth of a child often highlights important questions for parents about responsibility, self-identity, and notions of family, community, and society. For both fathers and mothers, the birth of a first child usually arouses intense emotions. For some, it is a time of fulfillment because they have created their own family. For others, a first child can arouse jealousy and feelings of inadequacy and leave both parents tired, confused, and feeling vulnerable, insecure, and rejected. The epiphanous nature of this kind of event may cause parents to reel into frustration and confusion or to gain a personal knowledge of what constitutes family, as well as a heightened awareness of themselves and their place in the world. At a more mundane but equally important level, the birth of a first child brings fundamental changes in parents' daily activities. It may heighten their need for neighborhood facilities, community services, social networks, and kinship ties. It may make them aware of certain difficulties with regard to the timing and spacing of these facilities, services, networks, and ties. At the very least, the birth of a first child will result in a transformation of gender roles and relations, and a renegotiation of the sexual division of space and time. The spatial nature of these changes and the fundamental questions that arise from them form a constellation of issues that define the diversity of family geographies.

The conversations with parents, grandparents, uncles, aunts, and friends that I record throughout the book are incomplete because my hearing is partial and my ability to write is constrained by my own understanding of family geographies. These constraints raise questions about my relationship to those we studied: How is my voice constituted? Whose voices should be heard? Who speaks for whom?

This book describes aspects of the production of space that arose from our conversations with household members, with particular emphasis on the power relations that are reproduced in gendered spaces. Although conversations were grounded in the everyday experiences of the participants, the communities they contrived, and the neighborhoods within which they found themselves, clearly those everyday experiences are influenced in part by large-scale material processes. My goal was to probe how the production of parental space cuts across ethnic, racial, gender, and class boundaries, and, to a certain extent, this goal was achieved. But the interpersonal nature of the fieldwork that the students and I embarked on limits how I can write about these larger processes even though our conversations with participants reveal much about ableism, racism, sexism, and capitalism. For example, class-based regional and global economic restructuring may explain, in part, the exploitation of labor or parents' differential access to resources, but commentaries on how late capitalism affects day-to-day life were rarely encountered in the field.

Representing what we found in the "field" embodies several other practical dilemmas. The work of James Clifford (1988), and George Marcus and Martin Fischer (1986) on what has come to be called "the crisis of representation" reveals that we rest our understandings of the world on precarious and unsteady foundations. As academics, we are no longer sure how to approach our subjects and how to write about them. We are even less clear about how to situate ourselves in what we study. A large part of this contemporary "crisis" reflects uncertainty about the methods we use to probe the lives and contexts of the participants in our studies. We now question how the relations "between researched and researcher inform our agendas and knowledge claims, how our work is affected by the communities and places we study, and how immersion in particular cultural (including economic and political) frameworks and academic and theoretical traditions informs research goals and methods" (Nast 1994, 54). Increasingly, social science researchers are questioning the space within which their research and writing is contextualized. Clifford Geertz (1983) points out that to have conversations with subjects that are distinct from everyday life we must have a field marked off in space and time. Although this site of inquiry is necessarily artificial and drawn by the researcher because of an assumed need for critical objectivity, the fieldworker usually experiences some form of displacement or peripheralization. The researcher goes to the field as an outsider and then draws on that experience to note differences and similarities between what is theoretically contrived and what appears in the field. Researchers, then, have heightened acuity that comes from their own displacement. This acuity is personal, self-reflexive, and, in the final analysis, it is what we write about.

Cindi Katz (1992, 1994) notes that our subject situation is composed of "spaces of betweenness," and the importance of this metaphor is that it highlights difference. As a researcher, writer, and educator, my work is always with others who are separate and different from myself. Difference is an important aspect of any social interaction that requires that we negotiate the worlds of self and other. Recognition of this difference leaves us always in-between, resisting definition and struggling for recognition. In the chapters that follow, I outline a theory for understanding the scale of difference between individuals, families, communities, and a larger polity. In the field, that theory impinges on my academic practice. In the play of difference in which we are engaged as researchers, Katz cautions that our movement should be outward to engage in the differences we encounter in ever-changing subjectivities and uneven power relations, but it should also be inward in an attempt to understand the differences and multiple subjectivities that constitute ourselves. In the words of D. W. Winnicott (1964, 1965), and to anticipate some of the arguments I make in chapter 5, we encounter in the field a "transitional space" where we are never "outsiders" or "insiders" in any absolute sense. With this work, I try to desensitize the dualism of self and other, but there is need to caution against celebrating the ambiguities and the disruption of any dualisms without questioning the extent to which power relations are upset by the ambiguity (Epstein and Straub 1991, 23). Power relations clearly exist in the field, as can be seen when informants want to give the "right answer" or in the subtle ways interviewers reflect their own political perspectives through body language and responses to questions. In addition, our power as researchers waxes and wanes as we vacillate between a clear understanding of what we encounter on the one hand and complete bewilderment on the other.

As I mentioned earlier, my initial place in the San Diego study as a passive observer became one of voyeur as I encountered my own difficulties with fatherhood. In time voyeurism was joined with empathy as I realized how many issues of parenthood transcend cultural, political, and economic barriers: my situation was similar to that of many participants. As a consequence, in talking of their contexts in this book, I give equal voice to my own frustrations, joys, and hopes. At the same time, many parents' life experiences and values are so different from mine that I could not find what Melissa Gilbert (1994) calls "mutual recognition." Often, I found the ways some parents met child-rearing challenges so upsetting that it was difficult to have a conversation. I take heed of Heidi Nast's (1994, 59) admonition that "camaraderie with everyone at all scales and levels of analysis presumes a kind of personal/political omnipotence, ignores time and space constraints that may be faced by both researcher and researched, and overlooks the fact that we may at times need to interact with those more powerful than ourselves or with those who blatantly oppress."

The early work of Clifford and Marcus highlights a politics of representation and the extent to which we, as researchers and writers, colonize the worlds about which we write. As Katz (1992, 496) notes, "Representation is a serious play of power in an overlapping set of historically and geographically determined social fields." How do we read, author, and represent space? An important extension of the work of Clifford and Marcus is the recognition that we cannot isolate academic writing from other kinds of social and spatial texts "without either overlooking the everyday contexts of speech acts or downplaying the great spatial and material diversities" (Nast 1994, 61) of those for whom we try to speak. As Mike Crang (1992, 541) points out, it is elitist to problematize our relations with others as simply a crisis of literary representation. We need to be able to recognize and represent other forms of social and spatial relations because from within them oppression, dissent, and difference may come to be recognized.

SITUATING THIS WORK

This book is not just about how I situated myself in the field but also about how my position was compromised and changed through that experience. I am now convinced of the importance of setting aside monolithic notions of the family, whether they be nuclear or New Age, so that we may address the politics of difference. Society tends to create myths of "ideal families" and "secure communities" that only constrain and shackle everyday lives. In reality, no norm can exist because we reproduce and reconstruct our family values and politics with every new event and challenge. In this book, I explore shifting family landscapes informed by my reading of feminist, geographic, postmodern, poststructural, and psychoanalytic theory. This reading intersects with information gleaned and created from interviews with the men and women who participated in the NSF project. For the most part, the book is a tribute to the people we talked with, who told us stories of their struggles with the day-to-day raising of children. Family fantasies defined and constituted by the raising of young children bias this work. All the stories contain elements of empowerment and commitment as well as of frustration and disappointment. Some of the mothers and fathers in our study are still together, while others define new family forms within separate households. Some of our participants are single mothers with "deadbeat" dads, and still others encompass themselves with an "extended family" of kith as well as kin. All are "brave new families" as Judith Stacey (1990) used the phrase when she described the day-to-day struggle of households trying to define themselves at a time when the myth of the family is de-centered and marginalized by increasingly volatile and unstable conceptions of space, place, and community.

There is an important geography here that goes beyond trite spatial metaphors describing "de-centered" and "marginalized" families. It is a geography of everyday life that contests definitions of space as simple two-dimensional mosaics reflecting patterns of social relations. This geography highlights the creation of a space that hierarchically produces, and is produced by, interdependent power relations. This space reproduces and reconstructs itself every day according to divisions of race, class, gender, ethnicity, and differential access to work and childcare. This book is about how these divisions embed families within particular ways of knowing and doing. The book is also about struggles to contest and break through constraining family fantasies and mythic ideas of community. It is about changing gender roles and relations, and new images of motherhood, fatherhood, and childhood. It is about attempts to unshackle ourselves from the discipline of our parents and the tutelage of their society. It is about embracing the good and beneficial legacies of our parents. It is about new lives that blow apart our reason and logic.

For most people, the myth of family life and community falls apart somewhere. This book is about that somewhere. It brings together theory from geography, planning, psychology, women's studies, and child development to address questions related to the power of families and the space of communities in contemporary Western society. My main thesis is that a series of myths that does not account for the diversity of day-to-day lived experience constrains our understanding of the spatial relations within families and between families and communities. For the most part, these myths encompass a monolithic family form that is only slightly removed from the nuclear family norm, and they encompass an idea of community that bears a strong resemblance to a mythic, small-town America. Although this nostalgia for the nuclear family and the small town is for a set of family and community relations that probably never existed, these images profess a set of emotive principles about how children should be raised. These principles constrain and contextualize a diverse constellation of people and places by superimposing a norm that is, for the most part, unattainable.

In chapter 1, I tease out this general thesis with a consideration of families as social and spatial constructions. I begin with examples from our study to highlight the unique and incomparable nature of contemporary family life. I then suggest that an evolving critical social theory of space enables us to increase our understanding of the implications of changing family power relations. We must understand these spatial power relations if we wish to focus on difference and diversity among and between families. In chapter 2, I explore some of the reasons why the modern monolithic family form gained such notoriety and why it continues as a powerful, pervasive fantasy in the face of diversity. The centrality of this chapter to the rest of the book derives from a discussion of the evolution of the mythic geographies and histories that contextualize families in communities and society. In many ways, these geographies and histories are necessary harbingers of our contemporary thinking about family form as described in chapter 1.

Recent feminist and geographic theories that attempt to unpack the myths and images that constrain our day-to-day lives inform a large part of the book. Chapter 3 focuses on images of motherhood and fatherhood as constructions and constructors of gendered family contexts. Feminist, poststructural, and psychoanalytic theories are woven throughout this chapter to help interpret some of our conversations with San

Diego families. Chapter 4 focuses specifically on how parents perform the work of raising children. The chapter identifies some of the limitations of understanding power relations within families in terms of the gender roles and relations of men and women. In this regard Judith Butler's (1993) notion of "gender performativity"—the performance of gender roles and relations—provides an interesting and appropriate framework for understanding the complexity of changing power relations within families.

With some understanding of fatherhood, motherhood, and changing gender roles and relations around the birth of a first child, I turn, in chapter 5, to a consideration of childhood and the question of justice in child rearing. The chapter raises questions about how certain roles that parents and children perform are legitimated while others are dismissed or remain hidden. Two theoretical propositions suggest a road forward. First, I propose that the "performances" of parenthood and childhood may find external legitimization in what Henri Lefebvre (1991) calls a "trial by space." Second, I expand on Winnicott's notion of transitional space as an intriguing way of conceptualizing justice for children (and adults). How spatial trials are continually played out in Western society—and how they relate to the myths of family and community—is the subject of the next three chapters. Chapter 5 ends by establishing the work of Lefebvre, Butler, and Winnicott as a theoretical fulcrum on which the book is balanced. In sum, I discuss the creation of a set of political identities around gendered family space on one side of the fulcrum (chapters 1 through 5), and, on the other side (chapters 6 through 8), I focus on how those identities produce, and are produced by, unnatural urban subspaces.

In chapter 6, I consider urban design and the development of autonomous suburban and private communities as contemporary reflections of the patriarchal bargain that continues to dominate residential urban space. I use examples of spatial entrapment and homework to highlight the ways middle-class women may resist the patriarchal bargain.

Communities are more than a spatial configuration of people; they also reflect, and are reflected by, ideologies and power relations. Chapter 7 unpacks some of the ideological debates that surround the myth of community, and chapter 8 opens the debate on the politics of differ-

ence to issues of justice and the scale of community. In chapter 7, I explore various contexts within which the myth of community is framed,
beginning with a comparison of the work of last century's fin-de-siècle
writers with the work of contemporary, postmodern, urban social theorists. The tension between fear for the safety of children and the supposed security of family-oriented communities is then explored. I
speculate on the ways that some women with special needs purposefully create communities of mutual aid and suggest that fear and security may be inappropriate precursors of connectedness and support. The
chapter ends with a consideration of the unnatural ways that families
and communities are sometimes connected in the academic literature
and how those connections often become the foundations of policy
change.

Concerns about justice and scale intersect with the topics of previous chapters, but in chapter 8 these issues converge. The chapter
builds a framework from which we can view the relations between
families and communities without constraining those relations to any
natural, hierarchical or linear notion of scale—without assuming, for
example, an ascendancy of power from the local to the global. The idea
of "jumping scale" suggests that some families are able to invent communities by creating, for example, a regional network of institutional
support. Regionalism may provide a flexible set of constructs for supporting difference and justice—constructs that are not necessarily tied
to what Judith Garber (1995) calls the claustrophobia of local "communities of place" or the elusiveness of achieving "communities of
choice."

Throughout the book, our conversations in the company of families with young children highlight the structural fragility and unnatural spaces of contemporary family life. The birth of a child, more than
many other events, simultaneously blows apart and reifies old images,
new hopes, commitments, responsibilities, feelings of self-worth, and
community identity. Changes in internal family power relations and
external community relations are highlighted when the commitments
of family members are strengthened, redefined, weakened, broken, or
abandoned as partners' interests change around the birth of a child. The
engagement with new parents throughout the book accentuates important contradictions between theories of space, community, and family,

on the one hand, and real-life experiences on the other. The confusion of family life I came across in the "field" cannot be disciplined into any one single paradigm or way of knowing from the "literature." My attempt to map family fantasies and community space is full of loose ends, circular arguments, and contradictions. Thankfully, our theorizing today is marked by a flexibility that enables me to embrace the contradictions and not worry about the loose ends. Ten years ago, I would have had difficulty writing a book of this kind.

Chapter 1

Re-placing *the* Family

A Space for Differences

For Doreen it all began to come together right there in North Park. "Oh, I don't know! North Park I like 'cause [of] . . . that neighborhood sense, and it makes me feel really safe and it makes me feel really comfortable. I don't know, it just reminds me of that old thing that they say that 'it takes a whole village to raise a child.' I like that sense of community."[1] This was my second interview with Doreen and many things had changed since our first meeting. For one thing, she had a son who was now six months old. She and her husband, Alonso, had separated two months earlier, although at the time of the second interview he still lived in the neighborhood and looked after his son on Tuesdays and Thursdays. Doreen had adjusted her work and school schedules to accommodate her new life as a single mother, and she had moved because of space needs. Throughout this turmoil, the community of North Park seemed to be an important and enduring factor for her. It epitomized a sense of place and continuity that, Doreen asserted, was rapidly disappearing in American life. "I just moved three blocks from where I used to live so if I take my son for a walk I still see the same neighbors. I like that community sense; it's a good feeling because a lot [of this] country has lost that. We're so locked in our cars and houses. Nobody has conversations anymore."

In many ways, North Park is an "urban village" as characterized by Herbert Gans (1962) when he coined the term in the 1960s. People in

North Park have "conversations" in the laundromats and grocery stores. You see people on the streets, and there seems to be an enhanced sense of familiarity and community identity. North Park is a compact neighborhood located about five miles northeast of downtown San Diego.[2] Crime levels are high and income levels are low in the area compared with San Diego as a whole. There are also more renters and fewer cars per capita. The neighborhood clusters around a small, mainly retail central business area comprising local grocery and hardware stores as well as coffee shops and bars. Given these superficial similarities, North Park differs in profound ways from Gans's "urban villages." For one thing, it is much more racially and economically mixed than the ethnic communities he studied in Boston. North Park does not feel homogeneous, organic, or natural, but it clearly offers a certain sense of community for Doreen.

Doreen's sense of community is not necessarily an existent reality: she did not simply find her community, nor is she quiescent among a supportive network of friends and acquaintances. Doreen actively works on the creation and maintenance of her community. "I'm kind of unique and an individual in that area of staying in a community; I don't think a lot of people share my idea." Doreen's community is in part created from her determination to build a sense of place, but it is also contextualized from the necessities of single motherhood. She waives her hand vaguely toward the open door of her rented, three-bedroom California bungalow:

> Yeah, well all my friends live in this neighborhood too. Actually [she laughs], I've been instrumental in getting all my friends to live in this area, but they really like living here because everything's here: you can walk to the post office; you can walk to the phone company. The women who get their nails done: there's a place to get your nails done. There's Jimbo's, you know, if you want to do that hippie food thing. And it's all right there!

Doreen's "extended family" differs from those described by Gans and his students because it is not maintained by blood or ethnic relations. Her local family is a purposefully constructed support network for herself and her son, Scott:

> I need my women friends. They are my family. My natural
> mother sends birthday and Christmas presents, and she visits;
> but it is to my friend Susan that Scott will run if he needs
> comforting and I'm not around. And for some reason I'd rather
> ask Nina to help out with Scott than my ex-husband. That
> probably just has to do with a boy/girl, husband/ex-husband /
> wife thing. I'm sure I will get over it, but I just feel like Nina
> would always be there—I've known her *forever*.

If Doreen's context constitutes a new kind of family geography, it is clearly a web of family and community relations that remains underrepresented in contemporary theorizing and planning. Feminist writers such as Elizabeth Wilson (1991) and Iris Marion Young (1990b) note the importance of high-density, service-accessible communities for single mothers, who increasingly constitute lower-income households. Their reaction is primarily to a duplicitous suburban ideal in which a patriarchal system favors the notion of a nuclear family norm.

A nuclear family was something that Russell and Trisha were in the process of creating when they first agreed to be part of our study. I was most impressed with their sensitivity to the quintessential geography of the nuclear family while at the same time being well aware of its biases and constraints. Already living in the peripheral urban community of Santee when Trisha got pregnant, they decided to move further out to a subdivision of large tract homes in Pine Valley. Santee and Pine Valley are primarily commuter settlements, but they maintain a rural, in some places redneck, sense. Trisha had quit her job a week before our first set of interviews in October 1993, and they had just moved into their Pine Valley home. Those two decisions pervaded our first set of conversations. Russell and Trisha shared their resolution to move to a more residential community than Santee and for Trisha to stay at home. At the first interview, however, Trisha seemed ambivalent. Russell commented that he "would like it if Trisha didn't have to work at all." Trisha talked of being both excited and scared about having a baby and about staying home all the time. "I don't know, I think I'm going to be bored crazy." A desire to find a safe place to raise children precipitated Trisha and Russell's move to Pine Valley, a small semirural community nestled among rolling hills. They also thought that buying a home would provide financial security. As Trisha put it, "We needed to buy a house. It's nice and it's quiet. And plus it wasn't close to everything: Santee is getting scary with the trolley coming in." Their house is a relatively large four-bedroom, two-story, single-family home. Several other young couples live on the street, which is wide and curves into another street that ends in a cul-de-sac.

Savannah was born in November 1993. By the time we organized the second set of interviews in July 1994, Russell and Trisha's struggle to create a nuclear-family haven was beginning to unravel. Russell was working over seventy hours a week to maintain the mortgage payments and spent considerable time during the interview bemoaning the California economy and expressing fears for his job. The toll on his fathering was also of concern. "The one thing I dislike about my job the most is the nights I work so late and [Savannah] is already in bed. When that goes on two or three nights in a row, it really gets to me." Trisha also felt that her mothering was compromised by their situation. "He works *all* the time! Sometimes I feel like a break, and there's no one to break me. I don't have family around, and I don't collaborate with the neighbors to co-op with day care or anything, so it is just me and her [Savannah]. Home alone!" Not only did Trisha feel isolated and cut off from support in her new home, but she also felt that the strength and independence she had enjoyed while in the world of paid employment were incompatible with being a full-time, stay-at-home mom.

In many ways, Trisha's position characterizes what feminist geographers call "suburban spatial entrapment." Some women's productive activities, they argue, are constrained by limited access to urban resources, by household and childcare responsibilities, and by patriarchal relations within the home. Trisha's comments on the lack of services and facilities in Pine Valley epitomize this thesis but also point to other underlying frustrations:

> Yeah, there's not community things around here. . . . Before it
> was so convenient; we lived in Santee, and everything was
> right, oh gosh, a three-minute drive to everything. Here the
> only convenient thing is [the] 7–11 [store]. . . . Like we were
> here moving in, and I had a couple of friends round, and they
> were like [ready] to get sodas or something. I jumped in the car,
> and I'm like I don't even know where the store is! So I had to

go and ask next door. It's trippy; we've moved out to Switzer-
land here.

A year later, Russell and Trisha's nuclear-family dream fell apart
with a trial separation and Trisha's move to her parents' home in San
Francisco. They are now back together and have moved to a more affordable house in a small community located sixty miles north of San
Diego. Both are in counseling and are searching for a creative way to
rebuild their family. In our third interview, at the beginning of 1996,
Trisha made these comments:

> The feeling of being out of control was that there was no such
> thing as a manual, and I was so used to that structure: this is the
> way to do it; follow this path and you'll be fine! If I choose to
> deviate there is no manual; and since I read I had so much
> information in my head. And I didn't have a support group
> down here. My family's up north, and I didn't rely on Russell's
> family because my ideals of how to treat an infant [are differ-
> ent], all the way down to basic things like should the TV be on,
> should the child watch TV, or should the music be loud. We
> were in conflict. . . . [Russell's family] always had the TV on.
> My life style's such that I didn't want to be like that. So I was
> very much on my own; and since I didn't have any guidelines
> that were cut and dried, I didn't feel adequate to make the
> decisions; and Russell was gone all the time, and so I took all
> my energy and all my everything and just focused on this child.
> I shut down completely from Russell. We wouldn't go out at all.
> Never! I never left her with a babysitter. She's never been with
> anyone who hasn't been in our family. I just let myself become
> overwhelmed with the fear. We live[d] out by Pine Valley, and I
> can remember taking the buggy [and] taking her out for a walk,
> and the overwhelming fear was that someone was going to
> come and abduct her. I was overwhelmed by the total responsi-
> bility [for] this helpless little child.

The family geographies created by Doreen, on the one hand, and
Trisha and Russell, on the other, are clearly unique and, in many ways,
incomparable. My intention with these conversations is not to look for
generalities or universal norms of family and community life. I do not

believe that Doreen's feelings of comfort and support in a relatively high-crime, racially mixed neighborhood have much to do with some structural component of urban living that produces "urban villages" or any other kind of geographic enclave. Nor do I believe that Russell and Trisha's decision to follow the dream of the nuclear family in a suburban or rural idyll necessarily caused any kind of breakdown in their family. Other stories from the families who participated in our study suggest the nuclear family ideal is alive, well, and working in Southern California. Rather, my intent with these introductory examples is to highlight the fragile and complex geography of family life and how it is constrained and contextualized by community space. There is a complex and important geography to family and community relations that has previously escaped much academic attention. This geography is much too alive and volatile to be considered a mere stage or setting for the idiosyncratic complexities of daily family life. The balance of this chapter considers the reassertion of space and scale in our thinking about social relations and why this focus can help us understand the politics of difference and the power of families.

Fragile Families in Unnatural Spaces

"In its broadest formulation, society is necessarily constructed spatially, and that fact—the spatial organization of society—makes a difference to how it works" (Massey 1994, 254). It is trite to say that ours is a spatial world. Nonetheless, the apparent naturalness of our everyday geographies until recently caused us to understand space as a mosaic that simply *contained* social activities. Neil Smith (1992, 60) points out that not until the late 1980s did geographical space emerge as a preferred concept for "interpreting" postmodern social experience. Since Jean Baudrillard's semiotic appraisal of the American psyche in *America* (1989), Fredric Jameson's (1984, 1992) exploration of "the cultural logic of late capitalism," and Henri Lefebvre's (1991) project on the production of space, academics have begun to reassert the importance of space in new and interesting ways. Among geographers, Edward Soja in *Postmodern Geographies* (1989) gives a particularly supportive account of the reassertion of space in the humanities and social sciences. Soja's work is an explicit effort to re-center geographical space "against the grain of an ontological historicism that has privileged the separate constitution" of time as the most important aspect of our changing lives (61). Although much has been written about how the family evolved in the past, how it is faring in the present, and what forms it may take in the future, little attention focuses on family spatial relations. An understanding of spatial relations is critical for unpacking the complexity of postmodern life. Taking his lead from Jameson (1984) and Michel Foucault (1977), Soja argues that contemporary culture is increasingly dominated by space and a spatial logic rather than by time. This logic is not linear or hierarchical, nor is it decipherable in any of our old ways of knowing.

The film texts that Jameson analyzes in the *Geopolitical Aesthetic* (1992), for example, are illustrations of the ways the postmodern world conflates ways of knowing "with geography and endlessly processes images of an unmappable system" (14). Put simply, our postmodern world is structured and reproduced by images with their own spatial logic, and this logic cannot be mapped by traditional methods of understanding. Jameson exposes the postmodern city as a living image that alienates people both politically and symbolically from new urban geographies. These new urban geographies comprise diversity and conflicts of scale that can, for example, bring together elements of the First World and the Third World in the space of a city block. As residents of this postmodern world, we are often constrained by old ideas and a nostalgia for the past that do not help us make sense of these new space and scale relations. Fuzzy images of white picket fences and reruns of *The Cosby Show* or *Leave It to Beaver* are conflated with ideas of how the family should be placed in an increasingly incomprehensible postmodern pastiche.

To begin unraveling the contemporary power of space, it is useful to describe how our thinking has changed since we viewed space as simply a two-dimensional mosaic occupied by social activities. Work in the social sciences and humanities throughout the 1970s suggested that space was neither natural nor merely a container of activities. With the work of Foucault (1977), among others, we began to understand that space was constituted through social relations and material social practices. The space of the home, for example, is constructed around a set of social relations among men, women, and children. These relations are hierarchical and reproduce power structures that are often based on a patriarchal system within which women and children are subordinate to men. No one should disturb dad while he is working in the garage or the study, for example, but the traditional domains of women, such as the kitchen, are used, walked through, and transgressed in myriad ways. Constituted in this way, space is a social construction that defines sets of power relations.

By the late 1980s, this notion of space also began to appear naive and one-sided. Lefebvre (1984, 1991) argued that the space-as-a-social-construction thesis implies that geographical forms and distributions are simple outcomes of power relations and material social practices. Lefebvre's ideas began to crystallize with the work of Jameson and Soja into an apprehension that not only is space a social construction, but the social is also a spatial construction. Soja (1985) coined the term *spatiality* to describe the dialectical processes whereby the spatial becomes the social and the social becomes the spatial (compare Pred 1986). Lefebvre's project is particularly important to this book because he not only unpacks the production of space but also suggests that concepts may become depoliticized and naturalized if they endure a "trial by space." The fantasies embodied within ideas of family and community have endured such a trial to the extent that they now reify a set of unworkable behaviors and practices in contemporary Western society. This idea relates to concepts like motherhood, fatherhood, and childhood, and the reproduction of culture through our children.

The evolution in our understanding of spatiality is perhaps best illustrated by changes in the questions asked of space and spatial relations. Initially, geographers were not much more than cartographers who mapped patterns and then asked, "Why do people do that there?" Questions were then raised to probe the social and cultural contexts of space: "What do people have to do to keep doing that there?" Finally, these questions were broadened to penetrate the power of space: "What is intrinsic about the way space is constituted there that enables people to keep doing that, and how, in turn, does that actively constitute space?"

Patriarchal relations not only are reflected in space but also are formed by space. Dad's garage and study, to return to the previous example, are often separated from the rest of the house by space, or at least a door. The kitchen, however, is usually the most accessible room in American houses and is often traveled through to get elsewhere. If mom controls the kitchen, does she also control the domestic sphere, or is "her place" one of subservience to some form of family norm? The power of space is equally evident on a larger urban scale when we reflect on the gender relations sanctioned by suburbs and the private, autonomous communities that are now so prevalent in the United States. Not only do we need to understand the ways in which meaning translates between the individual, household, community, and urban scales, but we must also develop a critical appreciation of the social construction of spatial scale itself. The evolving social theory of space begun with the work of Lefebvre, Jameson, and Soja is now joined by a critical appraisal of spatial scale (Smith 1992, 1993; Marston 1995). If we assume that power structures are reflected in, and formed by, space, then we must also take account of how they are hierarchically ordered. Smith (1992, 1993) and Jameson (1992) contribute most to our understanding of how societies are a product of scale. In addition, Sallie Marston (1995) interprets the production of scale from a feminist perspective. As an extension of Lefebvre's project on space, these authors point out that scale is neither natural nor incontrovertible.

"There is nothing ontologically given about the traditional division between home and locality, urban and regional, national and global scales. The differentiation of geographical scales establishes and is established through the geographical structure of social interactions" (Smith 1992, 73). Smith goes on to point out that the language of difference may well be articulated through spatial scale because the social construction of this hierarchical ordering creates borders and boundaries between people and places. My larger concerns, then, are with the formation of spatial power relations and how those relations are interpolated in Western society between individuals and families, families and communities, and communities and cities. I am concerned with the problems that arise from the metaphoric appropriation of the scale of "community" as something natural and its conflation with the monolithic and arbitrary values that are often assigned to the family. Such an appropriation enables society at large to embrace certain family and community forms and reject others. This broad indictment is focused in the chapters to follow around issues of spatial and scalar power relations. I ask why the behavioral and gendered contexts of the nuclear family are unworkable. At the same time, why do urban spaces remain incompatible with the needs of single parents, same-sex couples, and other families that deviate from the nuclear "norm"? We can embrace this paradox only if we accept that all families are spatially fragile and that our conception of the family is amazingly resilient in the face of diversity.

Fickle Family Forms

The formal academic study of the family is a comparatively recent phenomenon, coming first to conscious attention in the late nineteenth century, when rapid economic and spatial reorganization precipitated great social change. To illustrate how notions of the family are expressed in specific theoretical strategies, let me briefly sketch some interdependent and yet contradictory frameworks that often serve as templates for our understanding of families. These frameworks are discussed sufficiently in the family studies' literature so they need not be detailed here (see Bernardes 1985b; Gottlieb 1993; Stacey 1993; Popenoe 1993; Hareven 1994).

Some theorists uphold the family as a natural unit of biological and social reproduction that needs no sense of history or geography in order to perpetuate its structures. Adherents of this perspective testify to the naturalness and universality of the family, the biological imperative of the division of the sexes, and the rejection as "unnatural" of all variant family forms (Collier, Rosaldo, and Yanagisako 1982; Todd 1985, 1987; Durham 1991; Hey 1993). The modern family first began to look irrepressibly natural when it gained academic and institutional legitimacy with George Peter Murdock's (1949, 1) use of the term "nuclear family" to describe "a social group characterized by common residence, economic cooperation, and reproduction. It includes adults of both sexes, at least two of whom maintain a socially approved sexual relationship, and one or more children, own or adopted, of the sexually cohabiting adults." Murdock's definition is circumscribed primarily by how families function. This tradition of definition is derived from B. Malinkowski's (1913) structural functionalism, which asserts that the conjugal family is unassailable because it fulfills universal needs.[3]

The notion of the nuclear family is referred to as modern in contradistinction to the older notion of the "extended family." The extended family conjures up images of bustle, shared generational activities, diversity, and large size. The nuclear family differs from the extended family in that it comprises small, conjugal units of parents and children with restricted activities. Modern notions of the nuclear family restrict one part of its adult complement to wage employment and the other to domestic and child-rearing activities. Most often, these two occupations are filled by the adult male and female members of the family respectively, and they are thought of, at least in our enduring conception of the family, as mutually exclusive. According to Talcott Parsons (1955), the nuclear family has two other related characteristics: it is relatively isolated from extended kin, and it does not provide most of the former functions of families (such as health and education), which are taken over by other, more specialized institutions. With what is left of the modern family system, however, children might still be provided for, protected, nurtured, gendered, and molded until they too become adult participants of the system.

Second, another theoretical perspective on family form is encapsulated by Marxist social theorists, who are concerned with social and functional changes in the family that purportedly occurred during early capitalist society. Their perspective focuses on a historical evolution that mirrors supposedly predictable economic transformations. Early in this discourse, Friedrich Engels ([1845] 1968) questioned how the organization of production and reproduction on a societal scale changed reproduction and production within the household. Of particular concern to Marxists is the delineation of mechanisms that may lead from economic base to societal superstructure.[4] For example, how exactly does economic organization affect social reproduction? Within this framework, a particularly important family geography is contrived out of the distancing of home from work. In addition, Marxist social theorists speak about "family adjustments" to industrialization and about working-class "family strategies" for coping with a waged labor system (Hareven 1982; Glenn 1987).

Biological determinism is rejected by Marxists in favor of viewing the family as a social and economic construction and focusing mainly on struggles over the means of subsistence and production (Mackenzie

and Rose 1983; Swerdlow et al. 1989). Some believe that modern families no longer control their means of production, whereas preindustrial
households with their resources—usually land—had the advantage of
being the basic unit of economic and social production. Traditional
Marxist analysis suggests that new forms of family will evolve in the
future to liberate women and children, as changes occur in who controls the mode of production. The family here is portrayed as an agent,
independent of the individuals who constituted it. This theory assumes
that men, women, and children have the same goals and gain the same
benefits from maintaining a particular family form (Glenn 1987, 350).

Third, some theorists argue that any kind of structural-functional
definition of the family is reductionist and misses the complexity and
continual flux of day-to-day living and the place of individuals in family life. This "subjective" turn draws on humanism and clinical psychology to focus on meanings, identity, and individual definitions of family.
Notions of family are actively constructed as people interact with their
social and material world. Attention is directed to the possibility of
multiple family realities contingent on gender, age, race, ethnicity, and
citizenship. Two schools of thought on family evolve from this focus
on diversity. One sees the nuclear family as providing a core out of
which other family forms arise. Proponents of this theory point out that
although the nuclear family is incontrovertible, we need to appreciate
that everyone does not live in the same way. The other school of
thought contends that the scale of divergence from a monolithic notion of family is so vast that the very idea of the family is, and always
has been, redundant. It is to the debates around this second perspective that I want to turn now.

Theorizing Difference and Diversity among Families

Structures of gender, generation, race, and class result in widely varying experiences of what constitutes family. Historically and geographically, households vary in composition and function, and the relationship
between communities and families is too complex for most theories to
adequately prescribe. On the surface, the variation in contemporary
Western families seems to be the result of economic and social shifts
related to the steady rise in employment of women, high rates of di-

vorce, relatively low birthrates, increased life spans, and changing patterns of employment among both men and women. Some family historians would add that the complexity of day-to-day living in the past caused the family to be as disparately constituted as it appears to be now. Counter to the notion of the naturalness of the modern family, then, is the relatively new idea that the family is a redundant concept in the face of such diversity.

Jon Bernardes (1993) insists that the use of the term *family* instills the paradoxical compulsion to uphold "traditional family values" (compare Durham 1991). The perpetuation of this idea of family, as some feminists observe, results in enormous harm and the oppression of men, women, and children. Part of this oppression derives from people trying to make their own family more like the image of the family by striving to imitate a monolithic myth that comprises unworkable images of personal behavior, gender relations, and community embeddedness. Another part of the oppression is that the perpetuation of notions of the family also preserves inequalities in the geography of power, authority, and apparent wisdom.

Spatial power relations within the family changed significantly in the nineteenth century. The rise of modern spatial power structures accompanied the rise of industrial capitalist society, with its revolutionary social, spatial, and temporal reorganization of wage labor and domestic labor. In particular, two radical spatial and scalar innovations reconstituted family power structures: domestic labor and wage labor became separated, rendering women's work invisible and spatially isolated from the public sphere; and doctrines of privacy and territory emerged that attempted to withdraw family relationships, particularly those of the middle classes, from the community and from public scrutiny. Whereas the premodern family required extensive paternal involvement in child rearing, authority in child rearing was ceded to women in the modern era. From a spatial perspective this sexual division of labor found form at the start of the nineteenth century with the separation of the home from the workplace and the public from the private.

In a review of the history of American family decline, David Popenoe (1993) argues that the weakening of family power in society is not necessarily a bad thing. He claims (538) that the new "streamlined" family can now focus on its two most important functions: child

rearing and the provision to its members of affection and companionship. According to his framework, the family becomes the emotional
center of social life, and, further, the family is by far the best institution to carry out these emotional functions. His arguments are com-
pelling in the sense that he notes the importance of child rearing in a
nurturing environment, but they are also grounded in modernist notions of liberal individualism wherein the freedom of the individual is
the paramount value. The move from communalism to individualism
is often seen as the linchpin of change in modern Western ideology.
This is a precarious position from which to argue not only because it
implies some ideal historical progression of families but also because it
emphasizes "family types" and a technologically determined view of
"progress." The distinction between communalism and individualism
also suggests a problematic bipolar distinction between nostalgia for past
community and a "modern" liberal democratic belief in individual autonomy. For Popenoe, family power structures change as people disin-
vest in the family to pursue more spatially diverse, individualistic goals
of self-fulfillment rather than collective family, kinship, and place-based
community goals. He notes also that there has been an important
change away from patriarchy and that the decline in male power has
brought about female equality. He concludes that if these kinds of functional and structural changes are real, then we need only worry about
whether children are getting appropriate rearing and affection. But if
children are not receiving this care, then Popenoe's views imply that
the problem lies within the emotional structures of families. From
Popenoe's perspective, the problem with what remains of the traditional,
nuclear family relates to day-to-day complexities of family life only in
the sense that the family is unable to perform its core functions and
there is no alternative that can adequately do so either.

My distrust of Popenoe's argument stems from his institutional and
reductionist definition of the family, which consequently maintains its
core as a monolithic and natural concept. He neglects the elusive geography of family power that arises out of the mythic history of the family, the contradictions inherent in contemporary gendered ideologies
(and expectations), and the practical work of day-to-day parenting. For
one thing, Popenoe's assertion of female equality in the family is not
necessarily a reality. If parental space today is shared equally by mother

and father, some feminists would insist that the authority of the mother is less than that of the father and that many forms of patriarchy exist still. If so, then the day-to-day contexts and complexities of power and authority go beyond Popenoe's simple notions of egalitarianism. Alternatively, arguments from critical theorists such as Mark Poster (1978) and feminists like Judith Stacey (1990) and Dorothy Smith (1993) are appealing because they respect the power of emotion and the gendered complexity of family power structures.

Critical theorists and feminists argue against structural-functional definitions of the family from three perspectives. First, they question why the family is assumed to be a necessary condition for the survival and stability of any society. If, as a societal institution, the family is observed to be responsible for child rearing, does it mean that this function could not be performed if the family did not exist? To pose this question another way, does child rearing define the existence of families? Second, reducing the complexity and diversity of families to one monolithic family implies a progression in which some "family types" are more advanced than others. For example, the extended family is often seen as being inferior to, or more archaic than, the modern family. Third, the complex, day-to-day practice of family living offsets simple explanations of families that reduce them to powerless institutions that merely nurture and afford affection.

Some feminists and critical theorists argue that the family is the fundamental vehicle for creating and maintaining a patriarchal structure that oppresses women and children. Two interlocking structures of subordination both enmesh women and children in the domestic sphere with few opportunities for productive activities and socialize children to internalize male and female characteristics that maintain a patriarchal system of domination. Poster (1978, xvii) and Stacey (1993, 545) argue further that the family is not an institution but an ideological, symbolic, and emotional construct. The family is a universally legitimizing image, but families are multiple events predicated on their own cultures, politics, histories, and geographies.[5] A critical perspective positions family discourse not as a single culture, meaning, signification, or practice but as complexly interwoven social relations that connect the activities of actual individuals whose day-to-day contexts are "geographically and temporally disposed and institutionally various"

(Smith 1993, 51). In addition, because contemporary feminism and critical theory open windows on the gendered structure of the family and its emotional accouterments, a picture of difference and diversity emerges that not only makes it difficult to raise any one "family type" over another but suggests, more critically, the impossibility of family types. Diversity of, and within, families attests to the problems of suggesting any one monolithic family form. It also raises issues about our self-identity and our place in families and the tension between what we expect and how daily life might be constrained by those expectations if they are constituted within a broad family type.

Any work on contemporary families needs to move beneath and around specific notions of the family by focusing on the gendered spatial complexity of social production and reproduction. No social theory prior to modern feminism has satisfactorily explained why societies regulate reproduction as they do or how forms of reproduction are in themselves reproduced in the contexts of individual families. Feminist theory provides a starting point because it was the first to recognize that human production, reproduction, and consumption could not be studied separately. More important, this recognition involved an understanding of the relations between two sexes and between adults and children. Feminist geography enables a search for where the sites of reproduction can be contested, and, of course, early on it identified the home and the community as important and raised them above their taken-for-granted status. More recently, a conjoining of feminist and psychoanalytic theory with Marxism brings critical social theory to a point where we can look at the emotional and spatial structures that encompass the daily life of families and at the change in the responsibilities, feelings, emotional stability, and sexuality of household members. In later chapters, I outline how the evolving ideologies of motherhood, fatherhood, and childhood erect a broad emotional structure around contemporary family space. The next chapter sets the stage for this discussion by describing the mythic histories and geographies from which the idea of the modern monolithic family arose.

Chapter 2

Family Fantasies

Mythic Histories and Geographies

Over the last century or so we have retained a generally accepted image of the family in Western society. This image has saturated our collective consciousness, often serving as part of a common set of constructs with which we order and give meaning to the world. The image has remarkable power: it can be defined by most members of society, and it is supported and revered by many powerful political lobbies and policymaking institutions. A central assumption of this chapter is that the image of the family is, in some way, refined out of, and extended from, a complex set of historically and geographically mediated representations that are effectively hidden in contemporary discourses. Lefebvre (1991) uses the term "social imaginary" to signal the extension of a particular set of representations into the popular and collective imagination to the extent that it can be easily commodified and exploited. The family seems to fit Lefebvre's framework quite well: it is social because it implies societal rather than individualized processes, and imaginary rather than symbolic because it indicates a condition of possession rather than a state of signification (Ruddick 1996, 12). Given this societal possession, it is hardly surprising that the idea of the family continues as a problematically simplified collective and commodified cure for our social and environmental ills.

Why did the representation of a stable family form gain the status of a social imaginary? This is a heady topic, encompassing how a stable image emerged from changing ideas about family size, function, composition, and gendered complexity, and its political and economic integration within society as a whole. To focus this chapter, I concentrate on unpacking some of the changing spatial relations that contextualize our representation of the family and how it is embedded in community. I sketch some seemingly contradictory empirical arguments for evolution toward a modern family form, and I tackle the problems with assuming the historical and geographic immutability and naturalness of that form.

Reconciling the Social Imaginary of the Family

For many people, contemporary society offers little opportunity for communion, and so the social imaginary of the family provides a powerful anchor against seemingly hostile and alienating external forces. Given the powerful rhetoric that surrounds the notion of family values in American society, it is surprisingly unclear how they can or should be constituted. More critically, despite the moral forces that seemingly uphold family values in contemporary American society, many young fathers and mothers are disillusioned by the lack of societal support for the ways that they embrace the notion of family. For example, in an interview after the birth of her daughter, Cindy told us that she was "insulted that my work at home is devalued [by society]. I am frustrated by the fact that although society has paid it a lot of lip service, it still does not recognize someone staying at home to raise their children as a full-time job. I think this is obvious when you look at how society is not planned to support the family in its many different forms." Cindy gave up her career as a teacher to raise her daughter in what she considers to be a nuclear family setting. Her husband earns enough to support the family in an upper-middle-class life style in Rancho Cabrillo, a relatively exclusive, high-income community located about twelve miles north of downtown San Diego. In many ways, Cindy's context aligns quite well with the image of a modern family suffused with traditional values. If we engage for a moment the wisdom of Cindy's admonition that society supports neither her nuclear family nor the family "in its many different forms," the following two interview excerpts reveal a distinct dissonance between family needs and societal structures.

The first quotation is from a working mother; and the second is from Tatiana, a young, African American single mother who found support in an extended family.

> There's a lot of stigmatization given to a working mom, I think, especially from older folk. They want you home all day with your child. I'm very satisfied [with the balance between my work and family]. But it is the other, you know, like family, saying that you should stay home or hire somebody to come in and help you.

> I mean it would be really difficult if I didn't have my mom and, uh, my other grandmother [living with us]. I mean if I was out on my own, having to pay $500 a month in rent and everything, it would be quite a struggle. But, uh, as it's worked out, um, I've been able to provide well for [my daughter], and she's well taken care of, and she's well loved, and she's well spoiled. [My mom and my stepmom] just spoil Kirstin rotten, and my dad too. And then my grandmother that lives up there, she's crazy about Kirstin. She was always sitting around and complaining 'cause she's ninety-two. "Oh, I can't do this. Oh, I can't do that. Oh, I have this ache. Oh, I have that ache." And now all she can think about is Kirstin. So it's like given her a new spark to live for. Something to think about other than herself. So, Kirstin's got all these grandparents and everybody just dotes on her, so it's really worked out well for her. She's spoiled rotten [laughs]. She's got so many grandparents.

Tatiana lives in a run-down apartment complex in Chula Vista with her mother and grandmother. Chula Vista is a relatively low-income, mixed-race neighborhood located about seven miles south of downtown San Diego. According to Tatiana, there is a lot of gang activity on their street. Their apartment is on the ground floor so they always keep the shades drawn on a street covered with pot holes, broken glass, and graffiti. Although she tries to avoid contact with gang members, Tatiana notes that for many kids in the area "the gangs are family and security," providing protection that is not necessarily found elsewhere.

Politicians, community leaders, and academics often generate pleas to uphold traditional family values as natural, but what they mean by "tradition" and "value" fall into the rarely challenged realm of the social imaginary. Although many people are unwilling to vilify those whose families differ from the social imaginary, there is often an implicit bias against creating new family forms. One young mother articulated one aspect of the relationship between the values and the practice of the family social imaginary. "I think you need to be in your thirties to have [a family]. It's too difficult to be a single parent; it's just impossible. It's hard enough to do as a married individual [laughs]. I would never have had a child had I not been married. I just wouldn't do it. Just because it's difficult, not because of any moral issues."

Although the practice of parenting is not necessarily a moral issue, nonetheless, with their image of stability and security, normalized family values are frequently cast as existent and unassailable moral certainties. Given the seeming immiscibility of societal structures and the family practices represented in the excerpts above, it is surprising that the haven of the family remains pervasive as a basic assumption, or at least a goal, of many people. But if there is little societal support for the multiple realities of family life, there may nonetheless be generational support for the roles that men and women adopt in families even if these roles are sometimes at odds with the contemporary needs of families.

Living up to traditional family values was a significant source of stress for many of our participants. For example, this white, working-class, employed father turned to his father for help when he encountered difficulty coming to terms with his new family responsibilities:

> I was prepared for the responsibility of [a family], for knowing
> that we really wanted children, but the feelings [voice trails off].
> I knew I was going to be busy, but I thought it would be okay
> with a little extra effort [when I'm at home]. Well, I didn't really
> prepare myself for the stress of thinking about what's going on
> when I'm not there. I talked to my dad; he worked, and my mom
> stayed at home. So I knew I'd be working, so I talked to my dad
> about it. "Hey, that must have been difficult." I never realized
> there was so much stress behind it, and he sort of filled that in.
> That was kind of a big responsibility for him, but he kind of
> rationalized it in that he had a certain amount of pride by
> fending for the family kind of thing. That's what kept him going.

Generational experiences seem influential when individuals construct their image of how a family should be. Several men and women in our study referred to the image of the Cleaver family in the popular 1960s American TV sitcom *Leave It to Beaver* as an ideal of family life and one that they remember as similar to the image of their family of origin. Most recognize this image as a caricature, but it is, nonetheless, a powerful representation. The first quotation below suggests the bemused delight of a new father. This is followed by a quotation from Trisha, whose family I introduced in the previous chapter.

> I'll tell you a secret. I feel more and more like Ward Cleaver of *Leave it to Beaver* all the time. I do! I never imagined that being a father could be so satisfying. I really, really love my kid; I enjoy being a father. There's nothing I don't enjoy about it. I love putting her in a car seat; I love changing her diaper; I love feeding her; I love watching her watch TV. I just never imagined I could get this much satisfaction about being a father. I feel more complete as a person. And then there's the fact that I know I'm doing a better job than my father did, so [laughs] that makes me happy too [laughing]. Oh well, there you go!

> I was being so idealistic, almost Ward and June Cleaverish, that this is just how it is. And I left out so many gaping holes that now I have filled in quite a few of those holes. [The image of a nuclear family] is still very important, [but] I will do without a lot of things: I don't need a big house; I don't need new furniture, gold, jewels, cars. I don't need that, but I would like to stay home and raise her. That is very important to me.

Trisha's statement converges with the earlier quote from Cindy to suggest a seeming lack of support for the Cleaver family image in contemporary America. To state that the families in our study tried to model themselves after the Cleavers is a misrepresentation, but, taken together, these excerpts from our interviews suggest that the family evokes a distinct social imaginary. It is irresponsible to submit that this social imaginary is an outgrowth of 1950s' American family culture without careful consideration of some of the considerable historical and geographic baggage that contributes to a monolithic image of the family. Our image of family encompasses not only the familiar and comfortable nuclear family but also the idyllically represented endless folds of kin in the extended family and the complexity created when nonrelated people are also part of a household. The excerpts above highlight some of the implications of the reduction of this complex political and cultural reality to a monolithic family form, but they do not help uncover the hidden complexity. I shall bracket the dissonance between the stories of our young families and contemporary societal structures and put it to one side in order to focus on mythic family geographies and histories.[1] I highlight aspects of these histories and geographies to understand how the social imaginary of the family gained its hegemonic status.

Theorizing a Hegemonic Nuclear-Family Geography

For the most part, the suggestion that families changed in response to larger societal influences serves to reify the view of an emergent monolithic family form in modern times. Within this framework, families are viewed as a product of a single evolutionary trend, and by modern times family is writ large as *the* family (Glenn 1987, 349–50). Those who argue for this evolution posit important transformations of family forms in Western society beginning in about the twelfth century that related not only to population control but also to the need for security of wealth within family units, property-based individualism, and freedom in the marketplace. From this time on, parents no longer had to rely on numerous children or a family that extended beyond nearest kin for their security and welfare. According to this view, the size and cohesive solidarity of the extended family form were broken by the relentless pressures of modernization. The compact pattern of the nuclear family, comprising young people forming independent units of their own, enabled individuals to be more enterprising and less restricted by the tutelage and dependence of elderly parents than they had been in the past. These changes appeared to coincide with the beginnings of capitalism in England, then Germany and the rest of northwestern Europe. Capitalism and the industrial revolution were to spread these Western family values throughout the globe.

The notion of a linear development in family form—from the old extended system with its autocratic patriarch in premodern times to a revolutionary and progressive new form that spawned efficiency and individual autonomy—was popular until the early 1970s, when a group of Cambridge demographers headed by Peter Laslett suggested that the existence of an extended family in any period of history was a myth (Laslett and Wall 1972, Laslett 1977). Analyzing English parish records from the sixteenth century on, Laslett and his colleagues suggested that the family was stable in size. This work contributed to the view that an elemental family unit of parents and a few children is relatively unchanging not only through time but also over space. "The outstanding fact about nuclear-family households is that all through the past in the Western world, no matter where we look or how far back we go in time, they were extremely common" (Gottlieb 1993, 13).

Many contemporary scholars contend, in fact, that the nuclear-family household did not come into being as a result of modern capitalism but, rather, existed long before official records (Hey 1993). This sort of pattern was evident throughout Europe, with no consistent difference between urban and rural dwellers. Some family historians suggest that not only is this pattern remarkably consistent throughout Europe but any global variations are relatively easy to explain.[2] No empirical evidence, Beatrice Gottlieb (1993) contends, supports an extended-family norm, with several generations of married and blood relatives living under one roof. She dismisses the notion that contrasts in size distinguish preindustrial from modern households. For family historians such as Laslett and Gottlieb, the empirical evidence leads to one conclusion: the family is a natural phenomenon that seems impervious to historic or geographic inquiry.

In an extension of Laslett's work, Emmanuel Todd (1985, 1987) establishes the family as a natural infrastructure that determines the temperament and ideological structure of societies. Like Laslett, he considers the family to be a unit of biological and social reproduction that does not depend on history or geography to perpetuate its structures. The family, he alleges, is a blind, irrational mechanism whose power derives from its lack of consciousness and visibility; it is a natural anthropological form completely independent of economic and ecological environments (1985, 196). Todd generalizes this hypothesis to peasant societies throughout the world and to a corresponding variety of family types. In a lengthy series of anthropological descriptions of different geographic areas, he demonstrates how family types exist simultaneously in areas of differing climate, relief, geology, and economy, and, he points out, this geographic incoherence results in a nearly random distribution of family types around the globe. With Gottlieb, Todd concludes that the family is impervious not only to history but also to geography. The geography detailed by Todd is narrowly defined, however, as an investigation of the differences between places. Moreover, it does not account for differences across the life course of families or for differences in the spatial power structures within families. Todd's formulation seems more like descriptive cartography than like an unpacking of spatial relations, and, thus, he seems in a poor position to make universal claims about family forms.

The tendency to suggest universal truths in this way, to construct a monolithic family myth on scanty empirical evidence, is academically irresponsible. One problem with reconciling theories that are based on historical evidence is that the norm of the nuclear family had an undue ideological influence on past census takers, as it does today.[3] These compilers of information may have grouped people into ideal units rather than detailing actual complex combinations of units.[4] In a number of historical cases, the structure of the family appears to have been constantly changing and quite far removed from an elemental conjugal unit. Some researchers are now convinced that change, fluidity, and spatial heterogeneity characterized households then as today, and, importantly, we need theories to reflect this empirical complexity (Berkner 1972, 1975; Chaytor 1980; Wrightson 1982; Bernardes 1985a).

A further problem with deriving theory from historical evidence is that the family is most often defined by census takers through categories that reflect size and blood relations. These definitions do not enable theorizing beyond what the limited available quantitative data will allow. This form of theorizing, some scholars argue, is sorely inadequate because it cannot pose important questions that render family change intelligible from critical perspectives (Poster 1978). Recent evidence of difference and diversity across family histories and geographies should inspire academics to devise more critical theoretical approaches. The balance of this chapter adopts a critical perspective on changes that occurred in families and communities as capitalism developed.

Exploring Contingent Family Geographies

Theorizing from empirical data rarely encompasses family composition, the sexual division of labor, the evolution of the public and the private, and the spatial distancing of men, women, and children. This kind of history and geography of families finds its most eloquent expression in the work of Marxists, feminists, and other critical social theorists. Here, I begin by looking briefly at these critical questions in the early modern period; throughout I highlight the importance of family relations with community and the changing significances of scale.

THE CHANGING MEANING OF LABOR POWER AND PRODUCTION

Before capitalist development, Marxist social theorists argue, Western society was characterized by a unity in the production and reproduction of labor. The organization of labor was closely tied with the community, and the rhythms of production were regulated by need—that is, the amount of labor used in production was equivalent to the amount of labor needed for reproduction. Although most of their production was for subsistence, families produced some goods as commodities for sale or as recompense to the feudal lord for protection. In addition, families often bought necessary commodities in the marketplace. By the fifteenth century, this kind of serfdom persisted in a few places, but there was a steady transformation to money rents and sharecropping.

Nonetheless, Marxists speculate that the household controlled its own means of production at this time. Although there is little empirical evidence to suggest who controlled the means of production before the fifteenth century, we do know, for example, that in the middle of the fourteenth century, after plague had reduced the population of Britain by half, the sudden shortage of labor gave significant power to the working poor, who were able to free themselves from their feudal bonds. With feudal patronage behind them, and as the British cloth trade became important, the merchant class began to replace the landed gentry as a ruling class. In addition, rural households used their power over their labor to establish weaver cottages. This new "cottage industry" depended on woman power, but men were also drawn out of the fields and into the domestic domain. Along with children, they were set to spinning and weaving and the many other tasks of cloth production

(Robertson 1991, 108–11). By the time the scutching machine for processing wool was invented in Glasgow in the late eighteenth century,
most weavers were men (Smout 1969, 236).

NONRELATIVES AS PART OF THE SPHERE
OF FAMILY PRODUCTION

Not only did the household control the means of production in
premodern society, but it also usually comprised unrelated workers. For
example, at every level in British society in the sixteenth and seventeenth centuries except the poorest, children were expected to leave
their parental home at the age of puberty to become servants or apprentices. In the late-medieval and early-modern period the upper servants of a "great" household were drawn from the ranks of the gentry.
In the world of commerce any employee was called a servant; in the
countryside a servant could be an apprentice, a domestic servant, or a
farmworker. In general, the term "servant" was used to describe young,
unmarried women and men who lived with their employers and worked
according to the terms of an annual hiring contract (Hey 1993, 113).

However, servants were included within the family norm of the
time. The Latin root of the word "family" is *famulus*, or servant. When
"family" was first used in English in the fifteenth century, it was derived from *familia*, a collective term that encompassed only the domestic servants of a household. Only rarely was *familia* used for the entire
household, including the servants' employers. Gradually, the term broadened to include the whole household, and then, in the middle of the
seventeenth century, it narrowed again to its current definition as a
"group of related people" (Ayto 1990). Thus, the notions of household
and family tended to overlap before the seventeenth century, and they
came together in a spatially circumscribed unit made up of both relatives and nonrelatives. In this sense, the concept of family encompasses
an important geography.

This geography suggests that the servant in premodern times was primarily thought of, not as a paid worker, but as a member of the household,
a "unit of production" that coincided with a "unit of consumption." In
Marxist social theory, the production/reproduction/consumption triad
becomes quite complex, but it is undeniably tied to the spatial circum-
scription of family units. By the late seventeenth century, the familial

acceptance of apprentices, servants, and live-in staff had become much rarer in Western society as the control of the means of production moved away from the private sphere: in short, the geographical cohesiveness of the family ended.

The exclusion of servants and apprentices from the family and their demotion to "hired help" in wealthy families by the end of the seventeenth century had significant consequences for the emerging social imaginary. By the nineteenth century, nonrelatives were not recognized as being within the nurturing sphere of the family at any level of society. These changes began in wealthy households, where patriarchs had ambivalent feelings about those nonrelatives for whom they cared but from whom they also felt socially detached. This ambivalence eventually affected how everyone felt about servants; contractual labor began to take on new meanings and became increasingly relegated to the public sphere. Also, industrialization gave young people other work to do, and attitudes about the worth of the individual and personal autonomy created ambivalence among servants, whether they were low- or highborn. Usage of the term "servant" changed with the profound changes in social attitudes that arose, but only recently have we come to realize how important servants were to family life prior to that time. Changes in social attitudes toward servanthood not only reduced it to a low-status occupation and excluded it from the notion of the family but corresponded to changes in the geographies of families. I allude here to the beginnings of the creation of a public sphere that was spatially distanced from the familial, domestic sphere. Geography plays an important part in this evolution of the public sphere because it implies a change in spatial power relations.

SCALE AND SPATIAL CHANGES IN THE RE-CREATION OF COMMUNITIES

The boundaries between community and family, work and home, play and toil, and kith and kin may have been blurred in premodern times by the permeability of spatial and temporal relations. Spatial proximity and the embeddedness of work in home life and the community minimized the possibility of separate blocks of space and time for work, play, and family life. Although it is difficult to ascertain whether the means of production was in the hands of the lord or the farm household

in premodern times, the work of families was, for the most part, encompassed by the community. If we assume for a moment that there
were close ties between lords and tenants, whether through allegiance
or contract, and that the motives of the lord were not initially controlled by profit and tied to national and international markets, then
it is safe to assume that production, reproduction, and consumption were
controlled by the local community. In short, the farm household produced in conjunction with the community.

If we assume that change, fluidity, and spatial heterogeneity characterized premodern households, the considerable movement between
villages and hamlets made communities as transitory as families. This
mobility characterized not only servants going from job to job but also
households breaking up or settling in other villages in larger or smaller
landholdings acquired through purchase, inheritance, or new tenancy.
Urban neighborhoods offered even greater diversity and mobility than
their rural counterparts. In short, the permanence of village, hamlet,
and neighborhood populations in premodern times seems as mythic as
the notion of a permanent nuclear-family form.

Despite the transitory nature of communities, village life was important. The underlying ethic of communalism to which the majority
of rural households gave allegiance was threatened throughout Europe
from as early as the sixteenth century, however, as lords and landowners became enticed by international markets to turn common lands to
their own uses. For example, ownership of common land in Western
Europe, never an issue when subsistence practices were the norm, became significant when the lucrativeness of the international wool trade
persuaded many landlords to enclose large part of their estates for sheep
raising. This expropriation of land upset the village economy, jeopar-
dizing the productivity of moderately prosperous tenants as well as making life difficult for the large number of landless who were used to
foraging and gathering on the common land.[5]

Anthony Giddens (1990) argues that one of the most significant
consequences of crossing the threshold to modernity is the "disembed-
ding" or "distancing" of spheres of social life from the immediacies of
the here and now. Thus, the expropriation of common land in Western Europe for an incipient international wool market created a local-
global dialectic that simultaneously distanced landowners from crofters

and sharecroppers while producing a scale wherein the productive activities of tenant households could be appropriated and exploited. This does not mean that social life is no longer anchored in particular places, but these moorings are stretched from the beginnings of international capitalism across space. The "globalization" of social life with international capitalism is different from the territorial acquisition of empires because it occurs on a continuous and systematic basis (Gregory 1994, 118).

The village provided local cohesion and support for the rural household until the machinations of international capitalism evoked fluctuations between the local and global that village life was increasingly unable to accommodate. An important geography here relates to the rise of mercantilism and the creation of an urban ethos that was eventually to supersede the villalge-based communalism. Changes in settlement patterns with the evolution of a merchant society into capitalism had important implications for the spatial relations among production, reproduction, and consumption. As Derek Gregory (1994, 118) points out, the globalization of social life does not erase difference and spatial heterogeneity because local-global processes are thoroughly dialectic in that "events at one pole often produce divergent or contradictory outcomes at another." He goes on to note that with modernity local networks of social life may dissolve and recombine in a myriad of ways across an increasingly global space. Thus, with the rise of international capitalism in modern times, the geography of communities is transformed with the changing valence of space. I will return to this argument again at the end of the book because it is closely tied to how we understand community and why there is such nostalgia for "old" community forms.

PUBLIC PRODUCTION AND PRIVATE REPRODUCTION

In a study impressive for its breadth, Suzanne Mackenzie and Damaris Rose (1983) tackle the historic origins of the complex relations among social production, the circulation of commodities, and the reproduction of labor as they relate to spatial changes in families and communities from preindustrial to modern times. Following arguments that are now familiar in the feminist literature, they probe the importance of spatial relations in the separation of the "sphere of production" (work) and the "sphere of reproduction" (home). Their concern is that the spatial separateness of the domestic sphere is often considered to be at least an inevitable product of modern life, if not a natural product of life itself. They argue, from a Marxist position, that the history of this "separate sphere" has to be investigated in the context of people's struggles and contests around control over the means of production and subsistence because these struggles are integral to the development of the modern family during the transformations of industrial capitalism.[6] Their premises differ from those of family historians such as Todd and Gottlieb because Mackenzie and Rose see capitalism as creating the modern family form for its own ends. As a consequence, they construe this form as historically and geographically contingent rather than mythic and monolithic.

The capitalist organization of labor replaced much of the work once done in private households. Large numbers of unskilled and semiskilled workers, including women and children, were required in agriculture, mines, factories, and transportation industries. So-called white-collar workers were required in the modern banks, insurance firms, and department stores that financed, insured, and sold the new products of industrial society to the newly created consumer society. In all forms of this new social organization, people worked outside the home. Wages became the means of supporting the household rather than private family property in the form of land or small shops. Thus, the patriarchal preindustrial household lost its productive function—an important change that heralded the creation of the modern family form as described by Marxist social theorists (Swerdlow et al. 1989). These familiar arguments explain the great changes in family and community structure at the time, but they do not help us understand changes in spatial power structures.

In the beginning of this modern era, the function of the family in regard to productive, reproductive, and consumptive activities changed, but its form remained the same. According to Mackenzie and Rose (1983, 162), an inability to accommodate the changes caused by industrialization threatened to destroy the family. One of the ways that capitalist production expanded was to erode the rhythm of household activities by lengthening the hours men, women, and children engaged in waged employment. Increasing time spent in the productive sector left almost no time for the domestic sphere. The economic base of the preindustrial household had already been significantly eroded, but now capitalism was taking its toll on the human base. Long work hours in combination with low wages and, in many places, cramped and unhealthy living spaces enervated and demoralized the labor force. The household became a separate and private sphere where people pooled their wages to maintain themselves and where they lived. The household ceased to be the center of production, although it remained the center of family life. It was a sphere within which people could rest, eat, learn, love, and express feelings, but many of these activities were considered secondary to work: their timing, form, quality, quantity and, latterly, their geography became increasingly dependent on the relations family members had with the public sphere (Mackenzie 1988, 18). There was little time for the domestic production of goods and services necessary for the reproduction of labor because women and children were also at work in the factories. The notion of who makes up the family also changed at this time. The family shrank in size as it gradually lost the servants, apprentices, and live-in laborers who had dwelt within the household when it was a place of employment.

These changes in families created new problems for capitalism in the latter part of the eighteenth century. Capitalism was predicated on creating a new labor force from the preindustrial one, but that labor force was self-trained and self-regulated, and accustomed to an integration of living and working times and places. The new industrial labor force had to be disciplined, skilled in specific areas, healthy, and willing to work for a given number of hours every day. At first, there was an abundance of labor to serve the factory system in a relatively accepting manner, but these people were mainly unskilled and unused to machine work. In time, the precapitalist household, with its unity of production, reproduction, and consumption, collapsed and left a void because there was no mechanism through which a skilled and disciplined labor force could be reproduced. Meanwhile, high infant mortality rates and disease-ridden working-class residential areas left no doubt that the new household was unable to reproduce a healthy and literate labor force by itself. Not only was this a consideration for the expansion of capital, but workers—ill-fed, unhealthy, and often homeless— were becoming increasingly threatening as ideas of autonomy and individualism became popular.

I document these familiar processes to highlight the result of this discontent: the emergence of struggles for a separate domestic sphere, safe from the enervation of the factory and mine. Increasingly, philanthropists, private charities, and local governments intervened on behalf of the family to provide education and medical care. Initially, practical change began with the workplace: factory acts such as the Ten Hour Act in Britain in 1847 regulated working hours for children and women. Other acts controlled the working conditions for children by not allowing them to work at coal faces or to disentangle thread on cotton looms when they were motion. Still other acts required that children set aside a certain amount of time for education.

By the late nineteenth century the division of family labor began to take on a new spatial dimension as the place of women and children was recognized as being within the "safe haven" of the domestic sphere, cut off from the work life of adult males.[7] As more and more elements of family service (education, health) and manufacturing (processing food, making clothes) were transferred away from the household, the home became recognized as a separate sphere in which the labor force could be indoctrinated with appropriate values and attitudes of discipline and service. Working-class women and children still directed domestic labor toward producing some goods for use within the family, but these activities were increasingly being taken over by factories and institutions. A fundamental change was that domestic labor now had to reproduce workers, and unless someone in the household sold their labor power in exchange for a wage, the household would not be able to procure the goods necessary for survival.

The nascent power of separate domestic and public spheres lends credence to the notion that the social is spatially constructed and the disempowerment of family space is achieved through distancing it from the means of production. Crucial in the struggle to establish a new, modern family ideal in upper-, middle-, and working-class families was the growing notion that men needed to support their wives and children and that the means of that support came from a spatially separate public sphere. Concomitantly, private family life plays an important role in the stability of the social system: if the worker cannot get ahead socially, at least his children can; if work itself is alienating and exploiting, leisure time in the family compensates for it (Poster 1978, xviii).

In sum, the powerlessness of the early industrial family arose in large part because the home and workplace became economically and physically, then socially and emotionally, entrenched in separate spheres.

For some feminists the problem of powerlessness in early industrial families is constituted slightly differently as arising from the immiscibility of what was prescribed by prevailing gender ideologies and what actually took place in the public and private spheres. As more and more of the responsibilities of the preindustrial family were transferred to the factory or public institutions, women brought their traditional activities into the wage sector. By the mid-nineteenth century in Europe and a little later in North America, women numerically dominated the new public sectors of nursing and teaching young children. As factory work displaced the traditional, exclusively male-oriented craft manufacturing and as secretarial work began to replace male clerks, women were seen as displacing male workers. At the same time, women from middle-class and upper-class families became increasingly involved in charity and philanthropic work. Male exclusivity in the public sphere was further threatened by the demands of middle-class women for the vote, access to higher education, and unrestricted professional careers.

In terms of the prevailing gender ideology, women in the public sphere were seen as a threat to the newly constituted modern family. By the end of the nineteenth century the idea of women in the public sphere combined with declining marriage and fertility rates to suggest a breakdown of family ideals. In that families were still needed as a basic unit of consumption and for the reproduction of labor, the role of women in the public sphere was increasingly seen as unnatural and even dangerous. Organized labor unions' agitation for the "family wage" was based in part on a desire to keep married women in the home so that they might take care of the male wage earner and his children (Mackenzie and Rose 1983, 199). The family wage was in any case a myth because working-class women almost always had to contribute to household income. Most working-class women were employed in factories and the service sector, but, for others, an informal economy developed around activities such as sewing and mending clothes or being paid by wealthier families to "take in wash" or "child mind."

The family wage proffered one solution to what was dubbed the "woman problem." From the late nineteenth century onward, another solution began to evolve that was more closely tied to the spatial infrastructure of the city: urban renewal and the development of peripheral suburbs. Contemporary reformers saw the city, particularly slums and working-class neighborhoods, as reflecting and reinforcing the erosion of the family (compare Woodsworth [1911] 1972), and as a result the spatial separation of home from work took on a new, dramatic urgency. The causes of these changes are mixed and complex, but like the family wage, the disempowerment of women can be seen as part of the more insidious essence of suburbanization. This spatial disempowerment is so crucial to understanding family power structures that it is dealt with separately in Chapter 6.

The Emotional Heart of the Family

One problem with Marxist social theories of family change is the tendency to view the family as a dependent variable, a secondary structure that changes because of larger social, economic, and geographic forces. Marxists do not deal well with the issue of changing emotional patterns in families. According to David Harvey (1993, 23), one of the strengths of the Lefebvrian project on the production of space, and the reason many post-Marxists embrace it, is its unwavering commitment to the interdependence of materiality, representation, and imagination in that it denies the particular privileging of any one of these spheres over another. Another strength of Lefebvre's project is its insistence that only through the social and spatial practices of everyday life can the significance of all forms of activity be accounted for. On this day-to-day, practical scale emotions are registered and come into play.

An integral problem left unresolved by the Lefebvrian focus on space is changes in the conceptions of geographic scale that bear directly on our understanding of emotional patterns and changing family forms. A naturalized hierarchy of scales implies that people and places on one scale are nested within a higher scale, and each scale change implies moving toward greater abstraction and greater power. Thus, the individual is subsumed within the power relations of the family unit, and the family unit provides a unified voice for men, women, and children.[8] Similarly, the voice of the family unit is subsumed within the community, the community within the city, and so forth.

Researchers are in general agreement that the emotional values of individuals in premodern times were subordinated to family and community interests, and women and children ceded moral authority to the household's patriarchal head. These premodern family values survived civil revolutions in England, France, and the United States, but they did not survive maturing capitalism. As I noted earlier, the household was originally a productive unit that, through its labor, transformed various resources into food, clothing, and shelter. Initially, men and women shared these productive responsibilities, but with the development of mercantilism and the erosion of the premodern household, men became involved in the public sphere and women's emotional and moral labors became crucial to the survival of the household. In precapitalist society the household was the focus of economic and political relationships with other places and spheres of production, largely through kinship ties and networks propagated by women.

How did these spatial changes affect emotional and moral relations among household members and between households and communities? Changes in emotional and moral structures pertain to changes in power relations and control over the means of production. These transformations are difficult to decipher because it is never entirely clear which activities are part of production and which relate to reproduction, nor is it clear where these activities are performed. Feminist geographer Alison Hayford (1974) reasons that because women were functionally and emotionally central to the establishment and management of kinship ties in precapitalist times, with the emergence of capitalism it became important for them to be subjected to group control. This control helped to establish and maintain early capitalism. Economic, political, and symbolic forces often restricted women to the domestic sphere in precapitalist societies, but because the boundaries between public and private spheres were often unclear, women played a large part in family and community productive activities.[9] Hayford argues that although prevailing social conventions often demeaned women's labor, it was nonetheless essential.

In an evolving capitalist society, many of the geographically extended functions that were jointly accomplished by men and women, such as selling produce or managing a family firm, became part of an increasingly male-dominated public sphere. On the whole the eroding

core of the household was left under the care of women, as were the
moral values and emotional structures that underlay social reproduction. Authoritarian, premodern fathers who "saved the souls of obsti-
nate children by breaking their will" were superseded by virtuous and
moral mothers (Hareven 1994, 35). Some Marxists argue that, as the
public sphere grew, the private realm remained only because capitalism needed it to support those who could not work and to prevent their
dissatisfaction with the system.

As women were increasingly restricted to the domestic sphere, individualism became emotionally constructed in differential ways. Thus,
through the nineteenth and twentieth centuries, being a "father" became quite different from being a "mother," and the roles of sons and
daughters came to depend on age, sex, and consanguinity (Bernardes
1985b, 281). By the time Western society achieved industrialization,
there was a strong spatial component to the evolving identities of men,
women, and children, and they all became encompassed in the evolving social imaginary of the family. Separation of father from the private, domestic sphere became important for capital productivity in the
same way that the isolation of mothers and children established a mythic
image of the ideal mother who would guarantee both morally perfect
children and a morally desirable society (Bloch 1978; Marsh 1990).[10]
Fathers were bequeathed control of resources and productive activities
in the public sphere, and mothers were seen as having total control and
unlimited power over reproductive activities.

As a conduit into the discussion in the next chapter about how
conceptions of motherhood and fatherhood play out in the contemporary social imaginary of the family, I end this chapter by returning to
Trisha's image of the nuclear family. In many ways, Trisha's frustration
with trying to achieve a nuclear family illustrates some of the stress involved in attempting to live up to a normalized image of the family:

> Like just today the [family] counselor told us that the divorce
> rate was up to 60 percent. I mean, that's very, very scary, and
> you just see how fast it can [get to that point]! It was mostly my
> state of mind, [how I felt about motherhood and family]. I felt
> mostly that I was completely alone with the hours that Russell
> worked. I had a great income, but I still felt like a single parent,
> like a widow with a good pension. I mean my house was fine; I

didn't need anything, [but] I felt very alone and the whole entire responsibility: everything in the house, every single thing inside that house. You know, I remember filling out that checklist on your study's questionnaire of all the things that are your responsibility, and the only thing that I can remember not being my responsibility was the cars. That was overwhelming to me. And I remember checking it off and going "yeah, yeah," and now when you look at Russell's survey maybe he had most of his checked off too. And that was our lack of communication, but I remember just feeling pitiful. I was pitiful on myself. I don't know exactly what it was but, no, I didn't do anything that any other mom with a kid didn't have to do. I think that I took all the responsibility on myself, more being like a martyr. Some of it I enjoyed, and some of it I did not enjoy; and I think it was just lack of communication to say, "Hey, Russell, these silly things are bothering me." It is an interesting game, communication. But I need to open my mouth, and if he doesn't understand me, then I am willing [now] to say, "Can you sit here and [listen]," until he understands. But I would never do that before because I didn't see my mother when I was a child do any of that. She was June Cleaver! She just did everything. I never saw any interaction that wasn't benign. It was all very peaceful. Four children and [my dad's] kind of job, I'm sure they had conflicts. Well, I'm learning now how to deal with conflicts, and so it will be a little different; but I'm sure I'll still have the same responsibilities, but I think I will think them through a little differently.

Like all social imaginaries, the family endures in part through the maintenance of several myths, and the reification of these myths from generation to generation results in the kinds of day-to-day tensions outlined by Trisha. In this chapter, I have tried to highlight how these myths come into being, particularly those that support the reality of any spatial and temporal immutability in families. The purpose of this chapter was to tease out aspects of the mythic geographies and histories of families and communities as they relate to spatial and gendered power relations.

Chapter 3	Gendered Parental Space

The Social Construction of Mothers and Fathers

Gender identities and notions of the family are formed at birth and then molded by the events and circumstances that shape our lives. They comprise complex, reticulate, and multileveled frames on which we hang meaning and identity. Nonetheless, certain mythic structures may overshadow everyday meaning. These images contextualize and constrain our lives insofar as we feel compelled to play out the myths. Two quotations—from an expecting father and mother, respectively—are framed by specific myths of motherhood and fatherhood that emerge from what Deniz Kandiyotti (1988) calls the "patriarchal bargain."

> I make more money than my wife; if anyone is doing any adjusting [with this birth], it is her. It's kind of like the old traditional thing, the guy works and the mom raises the kids. Do I have a conflict between my work and my family? That's more of a chick question to me! I'm not the nineties sensitive guy. I'm the breadwinner. Will I get up with the kid at night? My wife has more of the homing instincts on that one. That's woman's work [laughs]. We pretty much delegate stuff. I'll do the car stuff, take out the trash. Very traditional.

> I think it's going to be wonderful [to stay at home with the new child]. I'm not starry-eyed. I think that there will be days when I'm bored, and I think there will be days when I think back and say, "Was I crazy to give up the security and prestige of the job I

had?" But in general I think it will be a fulfilling thing to do, taken as a whole. I'm really looking forward to having time where I can get centered again. I feel that I have lost my personal center in some way. . . . I'm going to look back in fifteen years and have a child that I want to be proud of, and my job is just not going to be that important to my life from a retrospective point of view.

This chapter revolves around the contemporary negotiation of the patriarchal bargain, which suggests that, within the modern nuclear-family form, women seek men's economic support and protection in return for domestic services and subordination. How are conceptions of motherhood and fatherhood socially constructed and spatially defined today? Some contemporary feminist writers argue that the controversy over monolithic family values obscures rather than illuminates the issues surrounding diverse, gendered emotional structures. They argue that the impulse of many families to imitate a monolithic nuclear-family myth is based on unworkable images of personal behavior and gender relations. The myth of the universal, loving family refuge, they point out, is punctured by the prevalence of domestic violence, and yet many women and children continue to endure abusive home environments. According to these writers, arguments over what constitutes family, both at the academic level and in everyday practice, are essentially over the patriarchal construction of motherhood and fatherhood.

The previous chapter ended with the suggestion that a focus on the emotional structure of families might enable us to understand how an ethic of individualism was constructed around the image of the family. In what ways are our contemporary social and spatial constructions of motherhood and fatherhood tied to an ethic of individualism and bolstered by the notion of a monolithic family form? This question was partially answered in the previous chapter with the suggestion that the spatial configuration of reproductive and productive activities through the nineteenth and twentieth centuries developed around a monolithic family form that separated out, individually and emotionally, our ideas of motherhood, fatherhood, and childhood. I now look more closely at how changes in families' emotional structures relate to contemporary gendered parental contexts.

Discussion in this chapter revolves around the formation of the monolithic gender categories of motherhood and fatherhood that are distilled from the multiple forms of masculinity and femininity. The mythic categories of motherhood and fatherhood are central to political-identity formation and social placement, but they do not reflect complex changes in patterns of daily living. Although many of the parents in our study thought of motherhood and fatherhood as "natural" categories that give meaning to their emotions and behaviors, further scrutiny suggests that they are complex social constructions with multiple dimensions and meanings that bear on families and other institutions.

(M)othering Theory

It may be argued that too much emphasis in our understanding of parenting focuses on the relationships between children and mothers. This focus may be justified in light of empirical research suggesting that increasing paternal involvement in childcare does not necessarily result in improved outcomes for children (Furstenberg 1988, 206). To put child-father relations aside for a moment, there is considerable concern among feminists that an overemphasis on child-mother relations has resulted at times throughout this century in the condemnation of women. Nineteenth- and twentieth-century myths about motherhood and Freudian psychological theory combined to create an environment in which mothers could be blamed for failings in children and, by extension, in society as a whole.

Linda Gordon (1992) writes that the exaltation of motherhood as women's chief vocation and centrally defining role is traceable from at least the eighteenth century. Traits of nurturance and self-sacrifice were used initially to argue the naturalness of the relationship between mother and child. In time, this relationship was defined as essentially maternal and, hence, womanly. Margaret Marsh (1990) notes that from the late nineteenth century these traits were given spatial symbolism in the home and community: idealized images of motherhood provided not only a basis for domestic design and planning but also a foundation for moral and family reform. A prescriptive literature began to emerge in the late eighteenth century on how to be a good mother. This literature focused on how to display appropriate affection toward the family and how to exercise proper moral influence at home and in the community—duties that in earlier centuries fell to the father. Throughout the nineteenth century in the United States and Western Europe, the bourgeois mother received moral training and guidance to enhance her motherly performance.

By the end of the nineteenth century the newly conceived science of home economics supposedly elevated women's domestic and motherly roles to the same level of importance as the public work of men. New household commodities appeared on the market to enhance the efficiency of domestic science. The geography of urban planning, particularly in the aesthetics and morphologies of suburban developments, reinforced the principle that a woman's place was, legitimately, in the home. As cleanliness became an issue in public health reform, new scientific standards of health became part of a woman's unwaged domestic responsibility. The definition of "a woman" and "the ideal mother" were conflated with the qualities of nurturance, emotion, and homemaking, and, in turn, these characteristics were excluded from the public world of work. As the home and women became constituted as the opposite of work and men, and as the writings of Sigmund Freud began to penetrate social thinking, mothers became "the other."

Consideration of the construction of the other suggests how mothering theory intersects with psychological theory. Nancy Chodorow (1974, 1978, 1989) was one of the first feminist psychologists to provide a coherent set of theories on motherhood. The self, according to basic Freudian theory, is what is left when a familiar object or person, most often the mother, is removed. As a consequence, much psychological theory posits mothers as other, the dark continent, the forbidden place of the oedipal myth. Noting Freud's focus on women as the other, Chodorow points out that the maternal is almost entirely absent from his account. She argues against Freud's notion of the other by suggesting that on a deep emotional level mothers experience daughters as less separate from themselves than sons. Mothers "mother" their female infants more than their male infants and thus reproduce mothering in females. Thus, in Chodorow's account, mothers' treatment of their daughters makes them both more dependent and more expressive than males. In effect, Chodorow naturalized and essentialized the category of womanhood, and, it can be argued, for this reason her early work became popular among feminist writers who were trying to establish a political identity for women (compare Wilson 1991).[1] Alice Rossi (1973), for example, stressed the natural and untutored quality of many women's intuitive responses to infants and the potential connection between sexual and maternal gratification. Many early feminist writers assumed that women were naturally the ones who should not only give birth to children but also handle primary child rearing plus performing nurturing functions for the whole society.

One of the more prominent of these writers, Adrienne Rich (1977, 282), speaks to the spatial and social isolation of mothering responsibilities that relates to the myth of suburbia and community. She argues that modern times diminished the necessary network of support for mothers and calls for its reinstatement:

> Various writers have called for a new matriarchalism; for the taking over by women of genetic technology; for the insistence on child-care as a political commitment by all members of a community or by all "child-free" women; communal child-raising; the return to a "village" concept of community in which children could be integrated into the adult life of work; the rearing of children in feminist enclaves to grow up free of gender-imprintment.

Radical feminist writers such as Rich wish to maintain a form of spatial isolation of mother and child in two ways. First, ideologically they want to make motherhood unique and respected socially. Second, they want practical conditions to change so that mothers receive the community support that would enable the specialness of the mothering relation to return. More recent feminist critique suggests that these emphases are misguided because the focus is on reversing the unequal power relations within the old patriarchal dualisms of work/home, public/private, and so forth, rather than subverting the dualisms themselves.

THE PRACTICE OF MOTHERHOOD

The isolation of mothers with children and the omnipotence of the mother-child relationship in family life are posed by Nancy Chodorow and Susan Contratto (1992) as a "supercharged" physical environment in which a fantasy of the "good mother" can be played out. Many of our respondents believed that the notion of the "good mother" was a natural trait of womanhood. As Doreen put it just after Scott was born, "It's just more of an introspective thing with him [the baby], and when I'm with him, I'm *really* with him. It's very natural. I'm not somewhere else, and I can just put my things aside."

If motherhood is perceived as "natural" by Doreen, such thinking also establishes a means through which she can maintain a level of self-esteem and create a gendered self-identity that helps her cope with a seemingly hectic and chaotic existence:

> I used to get really neurotic about work: if I couldn't go to work or I couldn't make this call or I couldn't do this, I'd get really really pissed and really really upset and just think [talking fast now], "How am I gonna do this? I need to be [gestures] blah, blah, blah! And you need to take the baby and do this." I don't do that anymore, because you know what? I've just gotten really calmer about it and everything works out and, really, there's just no accounting, or putting nothing on the same equality as raising a little human being. I mean they're *far* more important. Yeah, I like money, I like new things, I like to go places, I like to do things. But, you know, when I'm just hanging out with him and looking at him [pause], and what I'm doing is molding a human being that makes the world: it's like the biggest job there is! It's the most *thankless* job, we all know that [laughs], but it's a big job, and that's just put me all at ease with all the other bullshit that gets in the way. I just enjoy any of my time with him.

Some feminists are quick to criticize the patriarchal structures that constrain the potential of mothers like Doreen, but they also see a need to tackle the ideological basis of the "good mother." For example, some feel that if current limitations on mothers were eliminated, mothers would know naturally how to be good. Others are more concerned with a perceived need to reconstitute motherhood. For example, Eleanor Maccoby contests the naturalism and essentialism of Chodorow's assumptions with the suggestion that both boys and girls learn nurturing and maternal roles early on but boys are later constrained to deny this capacity (Maccoby and Martin 1983; Maccoby 1988, 1990). She notes that a sexist ideology gained social and academic legitimacy through essentializing and naturalizing women as mothers. For Maccoby, motherhood needs to be reconstituted as a social and political institution quite apart from biological determinism, and nonmotherhood needs to be accepted as a legitimate choice. This important reconstitution of motherhood transforms it into a less enervating endeavor that revolves around notions of mothering rather than motherhood. Motherhood wholly defines a woman for what she is; mothering defines her in a more limited way for what she does if she chooses to (Alanen 1994, 33). Some feminists embrace mothering as a framework that gives mothers a role as agents acting in society among other agents (Gieve 1987). Other feminists contend that mothering, no matter how it is constituted, is neither a natural nor an essential category of womanhood but is negotiated through the patriarchal bargain.

THE PATRIARCHAL BARGAIN

The patriarchal bargain, with its emphasis on what women give up, posits that mothers are not powerful but powerless under patriarchy. Some of this powerlessness is suggested from Trisha's discussion of her home life when she gave up her work to look after Savannah:

> Yeah, I don't transition well; and I had been in school since seventy-nine, and I graduated from San Diego State [University] in eighty-nine, and I thoroughly enjoyed myself. I enjoyed the structure. I enjoyed the [freedom]: I could take whatever class I wanted. I enjoyed the interaction of being told, "Okay, here's your parameters, and if you can stay within them and do them well, you'll get your reward." And that was a hard transition. I had my coaching and I had my job after I graduated, and so I was in control of everything. . . . But when I came home [with Savannah], I totally escaped into the house. All Saturdays and all Sundays when Russell had [time] off he'd watch the game, and I'd be be-grudgingly going through the house cleaning all the time. That's what I did, all the time! He would come home from work to *his* house, but for me my house was my home, my work, my relaxation. Everything I did right there! I never came home. And he'd say, "Why don't you just come sit on the couch?" And like I'd say, "Well somebody just

spilled popcorn and left their shoes and their socks and coat
and blah, blah, blah. And the house is a disaster and I cannot
relax." Whatever! It was very typical but very, very sad.

Jane Flax (1978, 1983) focuses on the difficulties of being a mother
in a male-dominated society and on the psychological conflicts and contradictions that such a setting generates. She asserts that it is time to
give up the myth of the all-powerful mother. "As long as we unconsciously believe that we can go home to her (or to nature) and she will
make everything all right (or that it is all her fault when things go
wrong), we can never reach maturity either as individuals or as a (potential) polity" (1983, 35). Flax also offers perceptive insights into the
contradictory needs of contemporary mothers. In particular, she feels
that women may unconsciously undermine their self-esteem when they
perceive a contradiction between nurturance at home and autonomy
in the workplace. For example, this young mother expressed frustration
about her dual identities as musician and mother:

> Freelancing as a musician, I often end up working a lot of
> nights as well, playing with the San Diego Symphony. It's the
> unpredictability that is a little tense. The baby has really, really
> cut into my practice time, and therefore at times I often will—
> not often, but once in a while—get in a really depressed mood
> and feeling like part of my life is being trashed out the door
> because I don't have time for myself and my baby. I love my
> work! It's fun and very gratifying. I'm nervous that [the baby]
> may not be getting [voice trails off]. I may not be giving enough
> of myself. For example, when I'm gone at night, I know he
> wants mommy. I'm just afraid he's going to go schitzo on me or
> something [laughs nervously].

Women's ambivalence toward work is reinforced by contemporary
work environments. Success for women often requires the suppression
of "female" attributes and capacities including childbearing and child
rearing even though the sexist ideology that prescribes many working
environments is beginning to erode as research unpacks the links between motherhood and female identity (Ireland 1993). Targeting essentialism in this way points to women's renegotiation of identity with

a greater appreciation of how women negotiate their roles as mothers. For instance, many women in our study resolved problems such as differential access to childcare by taking their children to work, but for this pharmaceutical worker childcare at home was equally a problem:

> Once in a while I take [the baby] to work and leave her [at the back of the pharmacy] in her car seat just sitting there looking at me [laughs]. I stuff her with crackers so she won't cry. . . . I think my husband should help more [when I get home]. There's been times when I say, "You take her 'cause I'm very tired," and he'll be like, "Oh, I'm tired too; I just got off work," and I'm like, "Okay, never mind, bye." He feels like that because, you know, like because he's at work he gets more tired than me: I shouldn't be tired 'cause I'm not doing his type of job; mine is more like using my head, not manual.

Some contemporary feminist research focuses on women's resistance to, and renegotiation of, the patriarchal bargain (for example, Dyck 1990; Marston and Saint-Germaine 1991) as well as emphasizing difference and diversity among mothers (McDowell 1991, Chodorow 1994). Still, we need to recognize that research into mothers as active agents who create viable lives within and apart from families does not diminish the effects of the patriarchal bargain. The patriarchal bargain still constrains mothers temporally by depriving them of opportunities for real economic power and spatially because the work of mothering is done in relative isolation.

Noticeable similarities and parallels in the social and spatial situations of women and children need to be investigated, but there is also a critical need to deconstruct the ideology and images of motherhood and childhood. I will argue in chapter 5 for a less constraining relationship between childhood and parenthood, but before that discussion can take place another actor at the patriarchal bargaining table needs to be introduced.

Fa(r)thering Theory

Even if we do not agree with the early work of Chodorow and others that the primary bond is between adult females and infants, we still need to question the role of fathers in family life. I noted in the previous chapter that the marginality of men within the family was established with the development of a property-based, industrial, and urban society. Since the beginning of the modern period, the father figure has become increasingly distanced from the emotional heart of family life. His image is associated with the office, the factory, and other places of waged labor that are spatially distinct from the home environment and the rearing of children. Fathers have been removed from the immediate geographical surroundings of children because of structural changes in the labor force over the last two hundred years and time spent at work.

An-Magritt Jensen (1994) argues that strong forces create even greater distance between children and fathers in contemporary Western society: with an increase in age at marriage and parenting, a decrease in number of children, and an increase in family breakups, children are an ever smaller portion of fathers' lives. In addition, in the United States today there is a growing tendency, amplified by the patriarchal bargain and economic necessity, for fathers to spend more time at work. The lengthening of work hours for men is particularly the case in households with young children. Our study of San Diego households reveals that for many fathers the amount of time spent at paid employment increases after the birth of a child, sometimes by as much as ten hours a week. During two separate interviews, Trisha voiced her frustration about the hours Russell works to maintain their suburban life style:

> I had no idea what to do with this little thing [the baby]: we moved, I quit my job, I had [a] Caesarean, and then my husband was working—wow—at least twelve hours a day six days a week. And that was too much for me to handle, and so I was very like "I have no control: I am completely out of control. . . . [Balancing] work and fathering is his issue. I enjoy a lot of time alone: I don't have a problem with that. I would have a big adjustment if he worked nine to five, five days a week. That would be hard on both of us. But these twelve-, thirteen-hour days, six days a week, and having [voice trails off]. He comes home saying literally, "Don't talk to me." He takes a shower; he sits on the couch and gets the remote; he says, "Hi Savannah"; and he turns [the TV] up and falls asleep

right after dinner. It is very frustrating, and I don't know what
to say. I figured, "Oh well, this is suburbia; this is how it goes:
he's a compulsive worker."

If fathers over the past century have been spending more and more
time away from home, we need a clear understanding of how this change
reflects on our ideas of manhood and fatherhood. These two terms are
not necessarily conflated in the same sense that womanhood was defined by the conjoining of motherhood and femininity so that motherhood came to define femininity. In many ways, in contemporary
Western society masculinity is separated from fatherhood with, as we
shall see in a moment, little understanding of fatherhood.

Some argue that there is an evolving, respected role for men as
nurturers and fathers. After all, men miss much by distancing themselves from their children. For example, it can be argued that the patriarchal bargain deprives fathers of concrete and continuous emotional
and spiritual bonds with their children. Mothering is often a deeply re-
warding, life-structuring activity that tends to create distinctive capacities for accepting responsibility, providing attentive care, and being
nonviolent (Ruddick 1992, 179). In the mid-1980s, proponents of a
model of a new, domesticated male produced optimistic figures suggesting fathers were spending more time with childcare activities (Pleck
1985, Lewis and Sussman 1986). Some studies of fatherhood suggest a
redefinition of gender roles with a particular emphasis on the paternal
contribution to childcare (Blau and Ferber 1985; Lamb 1987). These
studies note an increased commitment to domestic activities by males
in households where the mother is employed full-time and the father
is available to participate in childcare (Presser 1988). Michael Lamb
(1987) suggests that the emotional involvement of many fathers in the
development of their children has resulted in new and expanded definitions of fatherhood.

As a case in point, Allen is a self-defined househusband who de-
clares a great love for children. Allen lives with his wife, Janet, who is
employed full-time, and their daughter, Hannah, in a 1960s' home in
Santee, a city located about fifteen miles east of downtown San Diego.
Their home is situated in a small, working-class community nestled
against the mountains of a regional park. When I first spoke to Allen,

he had been his daughter's primary caregiver for nearly ten months. Feelings of nurturance came quite early to Allen. He told me how he left Hannah with the childcare provider across the road when she was a few months old, but she had difficulty napping and was "stressed out" when he picked her up in the afternoons. Because of this perceived stress and his general belief that professional day care is little more than "baby storage," Allen decided to take on the role of a stay-at-home dad, resigning from his job to look after Hannah despite the financial hardship. Allen likes his routine with Hannah: breakfast, one to two hours of reading, and then off to the zoo or a park for "running around." He likes to tire Hannah out because she then takes a long nap in the afternoon. When Hannah naps, Allen can get to the "domestic stuff like cleaning and yard work":

> [Hannah] is so good with everything. I haven't really seen her complain about anything, except when she's tired. She'll go back in her room and point to her bed, and I'll put her down. The only real trick is that if I really have to have something done and she needs to be with me, then I'll try to wear her out early and get a nap in before I have to go do what I have to do with her.

Janet contributes to the domestic sphere by doing all the cooking and most of the shopping. As she sums up, "Having Allen at home really, really helps. We found that it was more worth our while that he resigned from his job and stayed home. That just made all the difference in the world where I don't have to get up and get her ready, pack up the car or anything, drive her anywhere: Allen is definitely the primary caregiver."

When Hannah was one-and-a-half years old, Janet and Allen had a boy. Allen took on more of the cooking as well as looking after the two children, but he still preferred not to grocery shop. "Household responsibilities? That's me! Ya, everything except my supermarket phobia [laughs]. I just go in and buy stuff that I like. I'll come home and say, 'This looks good. Look at that price. I never even looked at that!' I'm looking at what looks good. I do my shopping at Seven-Eleven."

Allen did not think too much about how he was different from

other fathers he knew, but he did have this to say about being a
househusband, gender roles, and parenthood in general:

> Of course my friends talk to me about getting a real job
> [laughing]. Having all day to sit around and do nothing, though
> I'm sure they realize it's not just sitting around and doing
> nothing. Still, not having a deadline for much of anything! But
> I know personally that what I'm doing will pay off. I sometimes
> worry about just whether what I am doing is going to make that
> much of a difference, because even from a Brady Bunch
> situation you can get a child who can't cope [laughs]. You just
> never know. Hannah has turned out to be a perfect child so far.
> I just wonder what she'll be like when she gets older and
> interacts with other people and the social system at school. Just
> having two parents is probably more unusual [today] than just
> having a dad at home. The children who had parents that I
> worked with [at the special education school], most of the time
> it was one [parent]. If there was a mom, then there was no
> discipline at all; the kid just did whatever he wanted. And if it
> was just dad, then there was no emotion, just a real cold little
> kid.

Allen's description of the "emotional imprinting" of children suggests a theme in the patriarchal bargain that I will pick up on again in
chapter 5. He also made some interesting comments that bear on those
discussions because they relate to the naturalness of child rearing and
parent-child relations:

> [The kids] are pretty good at knowing when it's mealtime. You
> have to feed them, get them naps, and that kind of stuff, so it's,
> you know, a *natural* schedule as opposed to an *unnatural* one
> where you force yourself to start working at five in the morning
> digging a hole or whatever in construction. A lot of people I
> used to work with—especially in construction—think that I'm
> just sitting around all day and I'm watching TV every day. . . . I
> believe that kids have the same emotions and, and uh, actually
> a greater degree of concentration and ability to, uh, understand
> things than older people do. We build up little walls and
> prejudices and things like that, where a kid doesn't have that!
> And so, uh, you learn real quick whether you're just patronizing

them and pushing them away and, and paying attention
to them when *you* feel like it. If you're actually open to what
they're doing every time they need you, they learn that real
quick too and they become more or less receptive to being
annoying, you know! It works for me, but, again, I can't write
down or teach anything like that. It's just something I do
naturally. That's why I enjoy it.

Skeptics of paternal domestication counter that these traces of
nurturance and commitment to child rearing among men are observed
in only a relatively small number of cases, which may be inflated by
excessive media coverage (compare Roan 1991). David Eggebeen and
Peter Uhlenberg (1985) use U.S. demographic data to estimate a declining involvement of men in families between 1960 and 1980 of up
to 43 percent.[2] They interpret this decline to mean that the opportunity costs of fathering may be growing as the social pressure for men to
become parents declines. As I demonstrate in the next chapter, one of
the problems with these empirical studies is a lack of any clear idea of
what is meant by domestic and childcare responsibilities. Unfortunately,
the nebulous term *responsibility* now constitutes the emotional heart of
the debate about fatherhood.

GOOD DADS AND BAD DADS

The distinction between "good dads," who take part in child rearing,
and "bad dads," who shrink from that responsibility, leads Frank
Furstenberg (1988) to speculate that an ideological polarization is
emerging in the debate over fatherhood. Because of the externalization
of women's roles and their increased employment, he argues, men escape from "good-provider" responsibilities. At the same time, some statistics confirm that growing numbers of women are compelled to take
on the dual responsibility of motherhood and paid employment (compare Pleck 1985; Bielby and Bielby 1988; Lewis and Cooper 1988).
Mothers' economic and domestic responsibilities were highlighted by
Arlie Hochschild and Anne Machung (1989) when they coined the
phrase "the second shift" in a popular book of the same name to reflect
what women who are employed full-time do when they return home
from work. According to the 1990 U.S. Census, 57 percent of women
between the ages of sixteen and sixty-four were in the waged labor force.

Over three-fifths of all mothers were in the labor force at this time, and the majority (71 percent) were employed full-time (U.S. Bureau of the Census 1990).

Current estimates suggest that over half of all children growing up in the United States today will spend at least part of their childhood in a single-parent household, most likely headed by a woman. The single mothers in our study were particularly attuned to overt forms of patriarchy and commented forcefully on the lack of responsibility of the fathers of their children. Some of these so-called deadbeat dads left before the birth, while others left before the infant's first birthday. Many of these fathers did little more than support their children financially, and, for some, even this form of support seemed too much of a burden.

Those absent fathers who took on more than a financial burden were still perceived by most single mothers as not assuming their full parental responsibilities. Doreen puts it this way:

> Yeah I do, I really do [resent Scott's father]: I make more money than he does, so I've always had that going on; and that's a pressure, so I feel that and I have that! Even though he does pay some money, I've always paid more just because I make more money and it don't feel right now, even though he takes him every other weekend, as far as the day-to-day things like the medical care and the clothes he needs and the haircut—he needs a haircut [laughs and ruffles Scott's hair]—and just the basic stuff: *that* is all on me. I know what he needs: I know that he needs pajamas; I know that he needs shoes. 'Cause [his father] is not really here, so he doesn't see the stuff, even though if you look at what some other fathers do, he's pretty much there. Some fathers aren't there at all; you know they don't even pay any money. They bail out; and some, like this other guy I know, he's a father and he only sees his kids once a week. I don't think that's very much!

Some theorists suggest that fatherhood today has a voluntary dimension: fathers can either retreat from responsibility or participate in the family. Thus, although structural forces are widening the gap between children and men, forces at the individual level sometimes work in the direction of closing the gap. For example, here an expecting father shares his commitment to spending time with his child:

Some folks had to work to be able to have the things they
wanted to have, and they had to work really, really hard; and it
took a lot of their time—time that would be maybe an opportunity to spend with the kids and doing some different things
with the kids. And I think I want to try to [spend more time
with my kid]. So I will, even if it entails giving up some of the
extra work that I do, to spend more time with my child. I think
that's going to be an important thing. I know when I was growing up my dad was always, he was always working all the time.

RECONCILING THE POLITICS THAT PIT
FATHERING AGAINST MOTHERING

At the same time as the good dad/bad dad forces play themselves out,
some are outraged that judicial developments often favor children and
mothers over fathers. Thomas Laqueur (1992), a proponent for reconstituting fatherhood, is incensed not only that little is written about
what he calls "the new public man in private" but also by what he sees
as a return to naturalism in legal debates over "surrogate mothership."
This naturalism is indicated by arguments suggesting that motherhood
is a fact and, as such, it is an ontologically different category from fatherhood, which is an idea. He argues that although the fact of bearing a child is significant evidence, it should not be construed as "nature
speaking unproblematically to culture." Laqueur's concerns arise directly
from the increasingly challenged essentialist notion among feminists
that sex is biological while gender is social.[3] Although his primary focus is on the biological unnaturalness of motherhood and fatherhood,
Laqueur is particularly disturbed by the lack of a history (and geography) of fatherhood: "History has been written almost exclusively as the
history of men and therefore man-as-father has been subsumed under
the history of a pervasive patriarchy. . . . Fatherhood, insofar as it has
been thought about at all, has been regarded as a backwater of the dominant history of public power" (1992, 155).

Laqueur's concerns are not new. In the early 1980s, Michael Lamb
(1981) voiced concerns with feminist analysis that unpacked the ideology of patriarchy while ignoring the role of the patriarch. Similarly,
John Demos (1986) began an essay on the social history of fatherhood
by commenting that "fatherhood has a very long history, but virtually

no historians." Demos notes that although fatherhood, like motherhood, has had a checkered, nonlinear history, most change occurred in the twentieth century. Drawing from his own early work on family life in the Plymouth colony (Demos 1970), he suggests that fathers' roles in colonial America were closely related to supervising child development. Fathers tutored all their children in moral values from about the age of three. With industrialization, men's economic roles increasingly drew them away from the domestic sphere. At the same time, patriarchy changed as paternal authority over certain aspects of child rearing declined. The patriarchal bargain began to emerge. Control of resources from "elsewhere" made him a "good provider," but the new father was neither a pedagogue nor a moral overseer of the household, although he was still a disciplinarian. "A man's occupational standing established his authority in the home and his worthiness as a husband and father. This movement from ascription to achievement, which occurred throughout the nineteenth century, signaled a profound erosion in the role of fathers. And this transformation becomes one source of the good-father–bad-father complex that becomes more evident in the twentieth century" (Furstenberg 1988, 196).

Throughout the twentieth century, argues Furstenberg (1988, 199), a strict division of gender roles emerged that was precarious from the start because it set such rigid conditions for success. Ultimately, neither men nor women were willing to uphold their end of the patriarchal bargain. Furstenberg speculates that for men the breakdown of the patriarchal bargain is ultimately responsible for the rise of the good dad/ bad dad complex precisely because it gives fathers a choice apparently not afforded mothers.

Sara Ruddick (1992) counters feelings that we need to construct a history of fatherhood by suggesting that the story can never be positive. She too is concerned about the choices afforded men and women in families. Her typology of fatherhood, like Furstenberg's theory of good and bad dads, has two dimensions, but, for Ruddick, they are both negative: on the one hand, the story is of lost or gone fathers, irresponsible about providing cash or services to their wives and children, and, on the other hand, it is a story of fathers who provide as best they can but rarely assume a full share of the emotional work and responsibility of childcare. Mothers, by contrast—whatever their work, pleasure, or ambitions—can almost always be counted on to take up the responsibilities of "parenting." Ruddick's concern with the work of Furstenberg and Laqueur stems primarily from what she sees as both their unwillingness to engage with power relations in the family and their inattention to fathering as a kind of work.

Ruddick (1992, 181) raises the issues of power relations in the family in relation to "institutionalized father-love," which establishes fathers' claims to authority over children whether or not they have participated in the work of caring for them. She criticizes Laqueur because he seems to separate heart from hand (emotion from work) for all fathers, even those who try not to distance themselves from their families. Such detachment, she claims, leaves fathers with love for their children but little responsibility for the domestic labor that such emotions should rightly entail. Mothers are left not only with the emotions but also with the physical labor that has been historically placed on their shoulders. Ruddick suggests that only a focus on the work that childcare requires will rid us of the distinctions among the so-called physical, intellectual, and emotional activities of fatherhood and motherhood. She re-visions "mothering" rather than "parenting" as the work in which child-tending men and women engage. "Mothering," in Ruddick's eyes, becomes a gender-inclusive and therefore genderless activity.

The problem with our contemporary story of fatherhood, like that of motherhood, is that it is inseparable from the geography and history of capitalist society. Little is written or understood or problematized about fatherhood and domestic responsibility, whereas much more is understood about work separation, the productive capacities of men, and the power they wield over women and children. Such stories say nothing about "the new public man in private" or enable men to acknowledge or even imagine their domestic experience.

Victor Seidler (1995, 184) points out that men traditionally have great difficulty imagining the emotional space of child rearing, and their attempts are often fruitless because they have learned within a rationalist culture to deny that their emotions and feelings are a source of knowledge. It takes quite a different mapping, he goes on to point out, to appreciate that showing vulnerability does not have to be a sign of weakness. For example, here's what Russell had to say about re-visioning his role as a father in a re-created family with Trisha and Savannah just prior to their getting together again:

> I want to be there for my daughter, and I want her to see her parents in a healthy relationship. This is important to her. *All* the things that I didn't have when I was a child, and I didn't have a choice, she does! There are so many things I didn't know about having a healthy relationship. So, um, I went to counseling, and I really learned a lot, and, um, I'm leading a more balanced life, and [I am trying to stop being] such a workaholic so that I can dedicate more time with family, to myself. Educating myself more, which is really important. I had cool stuff and so forth and making eighty to ninety thousand dollars a year. That isn't so important. Um, before, when Trisha and Savannah—when we lived together—it's funny but it was never a concern. [Trisha] didn't work. She was able to stay home to take care of the baby. [When we get back together], maybe I'll work only forty hours a week; money won't be a concern. We won't save as much, but my life [voice trails off]. My life has changed. My life has changed a lot. I had *no* idea. A hell of a lesson. That's changing now.

Paradoxically, it seems that once forged and maintained in social contexts the bonds of mothers and fathers are often perceived as real and unbreakable. It is difficult to stop being a father or mother once you commit to having children. The power of essential and natural categories, such as motherhood and fatherhood are often assumed to be, may derive from their common association with attributes that are thought to be simple or natural traits, occurring prior to social construction. As Russell and Trisha's situation suggests, "naturalized" traits are difficult to change. Audrey Kobayashi (1994, 77) points out that we need to fundamentally challenge what we mean when we characterize "people according to criteria which may seem to present themselves as natural (because they have been naturalized) but are nonetheless based upon social choices." One way to question the naturalness of a category is to unpack its social, historical, and geographic components, as I have tried to do in this chapter and the previous one. In the next chapter, I look at practical, day-to-day changes in parents' commitments to childcare. I begin by unpacking gender roles and relations and then

attempt to weave our understanding of parental commitments and responsibilities around some of the concerns of Ruddick and Laqueur as
outlined in this chapter. I use data from the San Diego study to expand
Ruddick's interest in defining the work that childcare requires with an
analysis of how parents assume responsibilities and commitments after
the birth of a first child.

Chapter 4	Negotiating Gender Roles and Relations around the Birth of a Child

To what extent do parents renegotiate the patriarchal bargain through changes in gender roles and relations after the birth of a child? To answer this question, my interests in the social construction of motherhood and fatherhood in the previous chapter are joined in this chapter with a focus on how parents perform their gendered activities in the daily round. Ruddick's (1992) admonition that we need to understand the work of childcare in order to re-vision fatherhood and motherhood suggests a critical need to unpack the behaviors and responsibilities of parenthood. Understanding the involvement and bonding of mothers and fathers to their new infants is important because it relates to sexual politics and transformations in gender roles and relations. Seidler (1995, 183) points out that notions of liberal equality within sexual relationships often break down when babies are born: "Very little, within contemporary culture, seems to prepare young people for the impact that a small baby can have on a relationship and the kinds of dependencies that it creates." A family relationship that is carefully organized, in space and time, on the kind of rationalistic principles that underlie the patriarchal bargain may be unable to accommodate the changes that often accompany the birth of a child. But the San Diego study highlights a distinct lack of re-visioning around the birth of a child. Despite empirical evidence suggesting gender-role change, with men helping out more after the birth of a child, gender relations seem to remain circumscribed by the patriarchal bargain. In light of these results, Butler's (1993) notion of "gender performativity" helps us to rethink gender roles and relations in terms of parents' responsibilities and commitments. It also points to the possibility of understanding the ways space authorizes some performances while, at the same time, dismissing others.

Unpacking Gender Roles and Relations

Many contemporary mothers (and some fathers) seem to be oscillating between the poles of paid employment and domestic responsibilities with great voracity. Some empirical evidence suggests that in the United States not only are the perceived gender distinctions between men and women decreasing, but individuals' occupational and family attachments are becoming embedded in increasingly complex life styles (Bielby and Bielby 1988; Lewis and Cooper 1988; Wharton 1994). A longitudinal Scandinavian time-budget study noted that men in the 1980s were spending up to half an hour more per day with childcare than they did in the 1970s. Over the same time period, the proportion of fathers participating in childcare activities increased from 51 to 71 percent (cited in Jensen 1994, 71). Alternatively, some feminist research suggests that childcare shifts domestic labor more fully onto the mother's shoulders, with a consequent constriction of women's employment and leisure (Deem 1986; Michelson 1988; Loscocco and Robinson 1991). Although empirical work of this kind tends to garner data on gender-related attitudes while giving short shrift to actual behaviors, studies that focus on behavior through time-budget analyses unfortunately do not reveal any coherence on gender-role change either. For example, Joseph Pleck's (1985) analysis suggests that fathers spend substantially more time on domestic and childcare duties when the mothers are employed than when the mothers do not work outside the home, whereas William Michelson's (1986, 1988) analysis suggests that employed mothers still bear the brunt of domestic responsibilities.

One problem with these studies is the definition of actions and responsibilities. Lamb (1987) and Pleck (1985) draw important distinctions among behavior (interacting with children), being available (being around but not physically interacting with children), and responsibility (fulfilling long-term needs, such as making childcare arrangements).

Much of the increase in fathers' commitments revealed in empirical studies may be attributable to interacting with children with little change in long-term responsibilities. Some feminists insist that, even with changes in fathers' commitments to spending time around children, the authority of the mother is less than that of the father, and many forms of patriarchy will exist until men are forced to take long-term responsibility for childcare and domestic activities. If so, then the day-to-day enactment of power and responsibility needs to be considered carefully. Even if parental space today is shared equally by mother and father, the question of power relations within this space remains. A practical geography of family power will be elusive until we address the contradictions inherent in the gendered expectations and the practical work of parenting.

POWER AND RESPONSIBILITY

Early feminist concern with gender identity and behavior focused more on gender roles than on issues of power and responsibility. Gender roles are socially prescribed activities and forms of behavior. The complex patriarchal basis of gender roles is predicated on men's and women's adoption of these roles because of either a desire to conform or an unquestioned acceptance of them. The notion of gender roles is often conflated with essentialist constructions of male and female characteristics such as "providing for" and "nurturing." Women's unequal access to paid employment and their spatial entrapment in suburban neighborhoods, for example, are "explained" by their primary role as nurturers and mothers (Foord and Gregson 1986; Tivers, 1985, 1988; England 1993). Women's concentration in specific employment sectors has also been "explained" by gender-role characteristics such as assumed ability to do routine work, to be dexterous, and to be flexible (England 1993).

A significant aspect of contemporary feminist analyses is a focus on gender relations rather than gender roles within the family and the recognition of the significance of male power in these relations (Boys 1984; Croghan 1991). A focus not only on what people do (roles) but also on why they do it and how they feel about it (relations) moves research in a direction that does not mask the day-to-day complexities of family life. For example, some research suggests that when husbands participate in childcare, they are merely helping out and not taking responsibility for reproductive activities in the home (Michelson 1988; Croghan 1991; Wharton 1994). Doreen suggests here how complex such work can be while indicating at the same time an attention to detail that rarely surfaced among new fathers:

> The only thing I'm finding about preschools, the more research I've done in the community, I guess they assume that everybody goes to work at seven in the morning. Somebody needs to get some kind of preschool that's more flexible or that they realize that there's people that don't go to work that early, that [go to] work [at] ten or eleven. 'Cause even though they have flexible schedules, the flexible schedules are 6:30 A.M. to noon. Well I've trained my child not to see 6:30 A.M. ever in his life; we're not morning people [laughs]. I found one but it's way up in San Carlos. I'm not going to drive to San Carlos. [Interviewer: Have you found childcare in the neighborhood?] Not yet but I'm looking because I don't really, you know, I want to make sure it's licensed and registered. I'm kind of picky about that, I'm not going to drop him off anywhere.

A focus on gender relations can advance our understanding of why patriarchal power structures continue within families given the changes we have seen in family life since the mid-1970s. This understanding comes in part from the feminist critique of gender-role theory. Feminist geographers Jo Foord and Nicky Gregson (1986, 192) suggest that gender-role theory restricts understanding of peoples' active participation in creating social structures. Roles not only are socially constructed but also are created by people who have a choice. Thus, roles can be modified and changed by the varying degrees to which individuals either accept their existing form or are persuaded to adapt to new ones.

A focus solely on gender roles limits discussion to household activities and tasks without providing any insight into how men and women negotiate these activities. The following excerpt from an interview with a new father suggests the kinds of negotiations that occur between caregivers on a day-to-day basis and how these negotiations are translated into shared responsibilities:

> It takes about an hour to feed [the baby], and by that time she is ready to fall back asleep. [My wife] took the kid, and she started

feeding the baby; and I went into the kitchen. I was going to
make dinner, but then she said, "No, I'll make dinner"; so I sat
down and I started feeding the kid again. The kid eats pretty
frequently. That's what we're basing our lives upon right now,
when the kid eats. So my wife went in and fixed dinner, and I
sat down and fed the kid. I do a lot of feeding. She's been with
the kid the whole weekend while I've been at work. Being in
the Navy takes a lot of my time. I mean its not a nine-to-five
job, not at all, depending on how much work you have to do
and whether or not you can get it done all at once. I can go
home at noon some days, and I'll go home at six or seven on
others. If I'm home early, I'll make something up for us both.
We try to split up [caring for the baby] as evenly as possible.

A consideration of gender relations and responsibilities facilitates
an understanding of why men and women assume certain roles and not
others. Gender-role theory is a static form of social theory in that it
denies the importance of recognizing that these practices vary over time
and space. Teasing out local distinctions and personal dimensions may
lead to an increased understanding of the creation and persistence of
patriarchal power structures in terms of how men and women interact.

DIMENSIONS OF CHANGE

Changes in gender relations are in evidence when the commitments
of family members are redefined, weakened, broken, or abandoned as
the partners' interests change. Tensions result if we map onto the practical context of contemporary families some covert cultural norms of
modern family thought such as motherhood or fatherhood. The result
is what Stacey (1990) calls the "structural fragility" of family life, and
that fragility may be due to the lack of support provided by a political
and community environment that upholds the myths but contests the
reality of family living. Because of this lack of support from part of the
institutional structure, the unity of contemporary families depends almost entirely on the voluntary commitments of its members. This is a
contentious point that requires a more fully developed set of arguments,
which it will receive later in the book where discussion turns to how
young families create communities. The balance of this chapter discusses
how my understanding of family members' commitment and responsi-

bilities emerged from an analysis of the aggregate survey data of San Diego families before and after the birth of a child. The birth of a child, more than any other epiphanous event, stimulates the kind of changes that might highlight the structural fragility of family life.

To date, few studies compare domestic responsibilities and activities with differing levels of employment for both men and women. In addition, as far as I am aware, no one has tested suppositions about changing divisions of labor around the specific circumstance of the birth of a child. The bulk of the discussion that follows arises from analyses of returned mail questionnaires from households prior to the birth of a child and of a follow-up survey when the child was about six months old.[1] Both surveys focus on household members' day-to-day activities, responsibilities, and opinions about family life. The follow-up surveys enabled an assessment of changes in gender roles and relations after the birth of the child. In previous chapters, I discussed information gleaned from our in-depth interviews, but here I want to use the aggregate numerical data as the primary basis for discussion.

The research questions were designed simply to identify changes in the level of stress attached to various activities before and after the birth of a child. In figure 1, several broad categories of activities are ranked with respect to stress levels for pre-child women; care of infants is perceived to be the most relaxing and providing income is the least.[2] Prior to the birth of the child, both men and women rated all domestic activities except doing laundry, indoor cleaning, and washing dishes (and, for men, grocery shopping also) as relatively relaxing (a rating of 3 or less). All the activities in figure 1 became more stressful for men and women after the birth of the child, although this change is not statistically significant for any activity except one: men taking care of older children, presumably when the mother is occupied with the new infant. The activities that maintain the same level of stress are men's care of infants, grocery shopping, indoor cleaning, dishwashing, and provision of income for the household, and women's laundry and transportation of children.[3]

In a report of a study of stress associated with various activities among working and nonworking mothers and their husbands in Toronto, Michelson (1988) noted that perceived levels of stress may correspond to gender biases in responsibilities. Although Michelson did

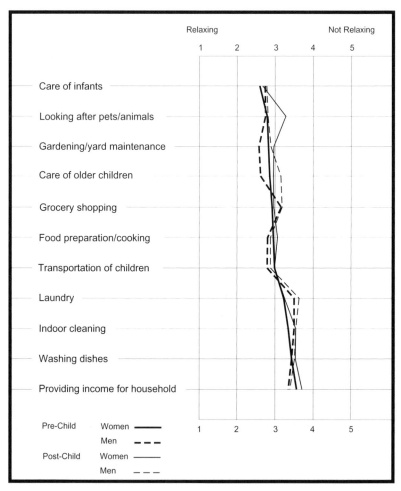

Figure 1. Level of Relaxation Associated with Broad Categories of Activities (Pre- and Post-Child)

not ask his respondents about their perceived responsibilities, his conjectures suggested an important dimension for our work.

The activities in figures 2, 3, and 4 are ranked according to women's perception of their responsibilities before the child was born. The figures document changes in women's perceived responsibilities after the birth of the child, and comparisons are made with how men rate the same activities. The major finding is aligned with most feminist writing in that domestic labor is women's work irrespective of employment

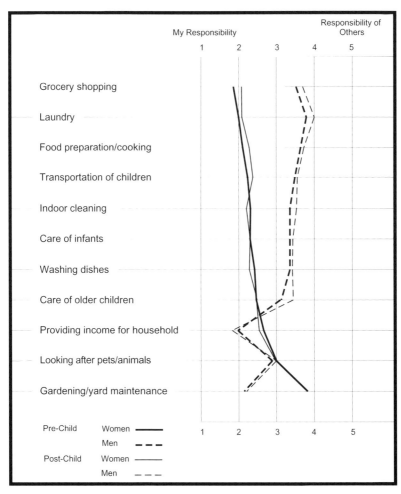

Figure 2. Perceived Responsibilities of Household Members with Full-Time Paid Employment (Pre- and Post-Child)

status or the birth of a child. On aggregate, post-child women feel responsible for nearly all domestic tasks except for gardening and yard maintenance. This gender gap in perceived responsibilities is statistically significant in all categories except "looking after pets/animals." Men's primary perceived responsibility is providing the household's income. Similar to the findings of Michelson's (1988) study of tension levels over a broad series of domestic activities, this relationship holds even when the women are employed full-time (see figures 2 and 3).

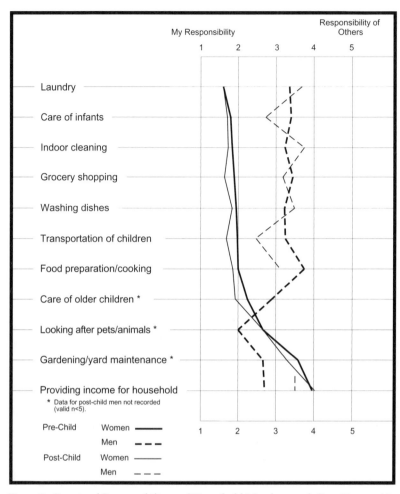

Figure 3. Perceived Responsibilities of Household Members with Part-Time or No Paid Employment (Pre- and Post-Child)

Figure 2 suggests that for employed men and women, the gender gap widens slightly in all categories after the birth of the child except for grocery shopping and food preparation. Mothers employed full-time seem to have greater responsibility after the birth of the child for laundry, indoor cleaning, the care of infants and older children, and washing dishes. The perceived responsibility of men is less for all domestic activities except gardening/yard maintenance. The largest slackening of men's responsibility is in the care of older children. Distinctions in

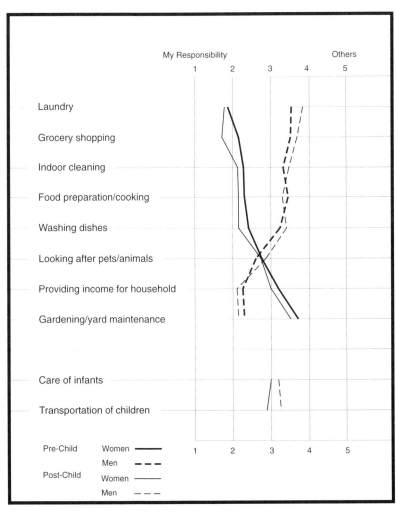

Figure 4. Perceived Responsibilities of Household Members with a First Child (Pre-and Post-Child)

pre-child perceived responsibility for providing income are reinforced at around the same levels, with both women and men indicating that they are more responsible for bringing in money to the family. Figure 3 suggests that although unemployed men and women, and those employed part-time, feel more responsibility for domestic activities and less responsibility for providing income than those who are employed full-time, the gaps between men and women are still quite wide. In fact, the gender division for domestic labor is slightly larger for unemployed men and women and those employed part-time.[4]

SOME TIME-SPACE IMPACTS OF A FIRST CHILD

On the assumption that changing gender relations after the birth of a child might be influenced by the age of the respondent and the number of children that a family already had, I set aside for separate study those who were expecting a first child.[5] I thought, somewhat naively, that younger men and first-time fathers might be more willing to buy into the media images of the 1990s' male and take on more responsibilities for childcare and domestic labor than older fathers.[6] Figure 4 reveals, however, the familiar story of a widening gender gap in perceived responsibilities for domestic labor. Again, for the most part, first-time mothers are shouldering more responsibility for domestic activities except gardening/yard maintenance. Men shoulder more responsibility for cooking and food preparation after the birth of the child, but it is still significantly less than that of the mother.

Time-budget diaries were procured from those in the first-child sample who agreed to be interviewed. From the diaries, it is possible to produce a pseudo-measure of commitment to roles by looking at the actual time allocated by household members to specific tasks. Time-budget studies have long been a useful way of collecting detailed information on the constraints and contexts of women's day-to-day lives (Tivers 1985, 1988; Michelson 1988; Rose 1993). Surprisingly, relatively few longitudinal time-budget studies address empirically the gendered division of labor (Hanson and Pratt 1988, 312). Information about daily activities was gathered in the San Diego study during in-depth interviews by asking participants to recall the activities of the preceding weekday. This information enabled us to construct graphs of time allocation (figure 5). As composites of participants' lives, these graphs hide as much as they reveal. In many ways, the representations in figure 5 leave men and women as disembodied aggregations. Nonetheless, the proportions represented by these composites tell a story that is not far removed from some of the stories I heard from the household members that we interviewed. At the very least, they give some idea of changes in time commitments after the birth of a first child. As expected, men are spending slightly more time in paid employment. Some men whom

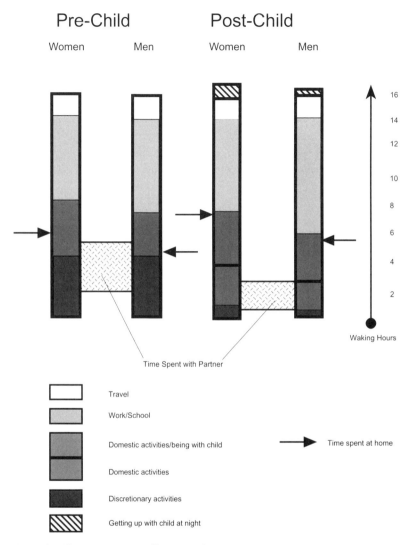

Figure 5. Changes in Time Allocations for First-Time Parents

we interviewed increased the numbers of hours that they worked by as much as 20 percent in order to maintain their families' life styles. On the whole, first-time mothers also spent more time in paid employment, but the increase was not as much as for their partners.

Adjusting to time constraints was a common theme among our participants during the first year of their child's life. Some first-time mothers

were clearly frustrated having to sacrifice their own needs to maintain
the baby's schedule; for others, coping involved acceptance of, and surrender to, the rhythms of a new way of life:

> My schedule has always lived for me, and now I really live
> around [my son's] schedule. I've noticed that [routine] is pretty
> beneficial to children. I think so far I've realized that if I live by
> his schedule, we all just get along a lot better. Last week we
> went shopping, and I just spaced out on the *time;* and he was
> getting really crabby and really hungry. I thought it was noon,
> and it was like 2 P.M. [laughs]. So I'm finding out that if I live
> pretty much by his schedule and he gets naps and his regular
> sleep, it works out pretty good. I'm pretty lucky, and he has got
> a pretty pat schedule; he does the same thing every day.

These new mothers sense that although schedule and habitual
routine are important, flexibility is also required. Here is how Doreen
put it:

> What I do sometimes is I break away from work at three [in the
> afternoon] and go to the library and study for a while; so I cut
> my income to study, and that's changed a lot. Or when I really
> would like to come home and relax after work, I have to hurry
> up and open up the books before he wakes up from his nap. So
> it's like I have to make every minute count for something;
> there's really not a lot of idle time to just piss around.

SURRENDERING TO TIME AND ENSLAVED BY SPACE

Lack of access to a car frustrates time management by limiting spatial
access. Previous studies point out that mothers face several challenges
because of transportation needs (Tivers 1985, Michelson 1986).
Michelson (1988), for example, notes that transportation causes more
tension than most routine activities because the tension is correlated
with lack of choice. Transportation problems for mothers are often acute
because the man's need for transportation to work usually takes prece-
dence. However, women's commitments frequently entail taking
children to a caretaker or school as well as getting themselves to work.
For the most part, the need for transportation is critical during the
highly pressured transitions between home and work. Time and space

collide for parents when they have to get children ready for day care and themselves ready for work.

Several of these issues came together for this new mother:

> [Being in several places and attending to several things at once] is a major concern. I used to have a schedule of 7:30 to 4, and I had to change to 8 to 4:30 because [my daughter] just didn't want to wake up in the morning. That is probably my biggest time problem right now, is that when she gets up in the morning, she doesn't want to wake up; she doesn't want to get dressed; she wants to do everything herself. So it's a big struggle. But if you let her pretend that she's doing it all herself, she's usually in a much better mood. But if your time is short, forget it. I deliberately chose a day care [center] that was closer to work rather than home so I can take the carpool lane on the way to work on the freeway.

Unique time-management strategies—such as this mother's choosing day care near work—are quite commonly used to offset time-space frustrations. Here is how Doreen expressed similar frustrations with time and transportation just after Alonso moved out of the house:

> When I [went] to school . . . when [my husband] was living here, I would just leave, go to school, and come back. And now I have to have everything ready, make sure [Scott] is fed, and pack him up to go. And after school, when I'm really tired and I just want to come home, I have to go—luckily [his dad] lives [close by]—pick him up and go through that whole [gestures with hands]; I can't just whisk him away. He's got to say good-bye to his father, [and I have to] be calm and cool about that.

Barbara's and Peter's lives changed significantly when their twins were born. As Barbara expressed her feelings, "Now I think twice before going somewhere. In the past, before I had kids, if I needed something, I'd run and do it. Now I wait until I have a group of things to do to make one trip. It's hard to take two kids with you. You never know what mood they're going to be in; they're on totally different, separate schedules, so they don't sleep the same times." Similar concerns were expressed by Peter. "Now it takes a little more planning with the kids.

When we decide to go somewhere with the kids it is not on a whim like we used to do."

Left unreflected in the aggregate statistics of figure 5 are particular events and contexts that may cause space-time stress. Alice is a single, African American mother whose child, Mary-Jane, was born with cerebral palsy. She describes the problem of getting to her daughter's hospital appointments when she lost her car. "Yeah, there was a time when I was always pushed for time and stressed out. I lost my car; so for a long time I took the baby on the bus. I was working an eight-hour day; I had no roommate, a baby who was sick; I had to get to doctors' appointments on a regular basis (sometimes twice a week), so things were hectic." Alice goes on to voice her frustration with public transportation:

> There is no bus that goes down Texas Street. I was appalled! It
> took me two buses to get eight blocks away. I would take two
> buses there to get Mary-Jane [to day care] and then two buses to
> work, then two buses to pick her up and two buses home. And
> then I found I could walk to work. I would take the bus to drop
> her off—because, you know, she's heavy—and be at work five
> minutes faster than if I took two buses. And so that was just my
> morning exercise. I made the best of it.

For some women, surrender and acceptance are simply a part of life with their new child and a way to cope with time and space constraints. Importantly, making the best of it can be an uplifting and positive experience. Alice explains that she often misses taking the bus with Mary-Jane now that she has a new used car:

> [Mary-Jane's] been great for me! I got more accomplished with
> her. Even on the bus: I got so good with her, you know, if I had
> to be someplace in ten minutes we'd be packed and out the
> door and at the bus stop. We got so good on the bus, everything
> packed up. With driving in the car, oh gosh, it takes a lot
> longer to get going because there's just a whole [lot of stuff], and
> always several other hitches in the get-along. A lot of times
> with the car I just think, "A quick stop for milk. No, can't do
> that!" So a lot of times in the car you're not as focused; you
> don't get as much done. On the bus, you don't do any side trips
> on the bus. You go and you get there and you come home! One
> of the things about being on the bus is that you get to hold your

baby; you're spending a lot more time . . . with your baby. In the car you can't hold them. You can't snuggle; it cuts down on your snuggle time. I enjoyed [the bus] because Mary-Jane was my traveling companion; we had fun. We got out together.

THE WORK OF PARENTING

For some women, acceptance and making the best of it with their children are not options that they prefer. But they find that infant-oriented spatial and temporal constraints do not necessarily curtail the activities they feel are essential to their own well-being if they find appropriate support. Here is what Barbara had to say about having twins:

> I'd do it all again! I told Peter that if I could have another set of twins, I would! I'd have two sets because I always wanted four kids. Peter's wanted three, and I've never wanted three. So I said if I could guarantee me another set I'd do it; and he says I'm absolutely nuts [laughs]. He says that's why you want them, because you leave and every other week you're on the road, so you wouldn't have to take care of them half the time [laughing]. He's a great dad; he really is. I couldn't do it without him.

Figure 5 suggests that fathers in the San Diego study are committing more time to domestic activities (including being with the infant) than they did before the baby was born. Our discussions with parents like Barbara and Peter do more than bolster this impression; they suggest that the mothers in these families would have great difficulty coping without this time commitment from their partners. For example, the following two excerpts (Barbara first, then a young father) suggest a reciprocity that was sometimes voiced by new parents:

> They [the twins] usually wake up and drink a bottle and go right back to sleep. So I didn't have to do anything with him. After he was fed, he went right back to sleep. [Interviewer: Do you normally get up with the kids?] We trade off. It all depends. Lately, we were on vacation last week, and the last couple of weeks he's been getting up; he was cutting a tooth so we share as to whose turn it is. Other than that, up until this point, they've been real good at sleeping through the night, but we usually take turns if someone gets up late at night.

[The baby] went to bed around eleven and my wife around
twelve. I went to sleep around 1:30 A.M.; the baby was crying so
I went there and I held him and I realized he was hungry; so I
woke my wife and I told her about it. And around six or
something in the morning the baby was crying, and my wife
told me, "You should do your father duty" [laughs], so I tried to
pacify him for an hour.

But the complex changes accompanying the birth of a new child
can, in many ways, conflate overall responsibility and commitment with
merely helping out at a day-to-day level. The apparent changes in ideologically based responsibility and practical commitments around the
birth of a child are explained in the literature in a variety of ways. For
example, in a study of the attitudes of young adults toward parenthood,
Philip Morgan and Linda Waite (1987) conclude that couples become
more "traditional" in their gender-role attitudes with the birth of a first
child. Similarly, Frances Goldscheider and Linda Waite (1991, 138–39)
conclude a report of a national study of U.S. parents with the assertion
that couples find themselves becoming increasingly less egalitarian when
children are born:

> Before children are born there are fewer tasks (and perhaps
> time together is particularly valuable); [couples] can be "true"
> to their egalitarian ideals. When they are young, idealistic, and
> egalitarian, they can start off sharing, wrapped up in each other
> and their common tasks. But such couples usually find them-
> selves becoming less egalitarian when children are born, since
> they can no longer afford the time to shop or cook or run
> errands together—one must watch the infant while the other
> accomplishes the needed tasks. And at the same time, the
> family is under increased economic pressure, since children cost
> money, [which] . . . puts added pressure on young fathers to
> work longer hours and work harder. Hence, the diminishing
> hours husbands have to spend with their families become more
> necessary for rest and the release of tension. . . . As time
> becomes rare, and tasks more pressing . . . specialization of tasks
> by gender becomes more efficient

Although parts of the San Diego data seem to corroborate this
speculation, I find functionalist assertions such as these to be extremely

problematic because they do not call into question how gender roles and relations are emotionally constructed. Seidler (1995, 183) points out a possible critical contradiction between the rational and egalitarian basis of many young couples' sexual and political identities prior to the birth of a child, on the one hand, and the irrationality and emotion of the day-to-day work of childcare on the other. To return to Ruddick's (1992) concerns, outlined in the previous chapter, it is important to try to envision fathering and mothering as a kind of work. This work is not so much about sharing domestic responsibilities, although I am completely sympathetic with arguments for gender-role equality. More important, the work of parenting requires different kinds of thinking, involving not only the practicalities of family economics but also passionate emotion and rigorous honesty about feelings. It raises issues of gendered identity for both parent and child that need further exploration. It exposes the concepts of fatherhood, motherhood, and childhood as unnatural, contrived, and delicate. Parental bonds with children should not be viewed as natural and factual but as constructed and flexible. In the next chapter, I try to tease out notions of child rearing and parental control around some of the gendered contexts raised in this chapter. To set the stage for that discussion, I conclude this chapter by interpreting the findings from the San Diego study in relation to how gender roles and relations are performed as part of the patriarchal bargain.

Gender Performativity and the Patriarchal Bargain

Butler (1992, 1993) uses the term "gender performativity" to encompass the day-to-day performance of gender roles and relations. Gender performativity cannot be understood "as a singular or deliberate act, but, rather, as a reiterative and citational practice by which discourse produces the effects that it names" (Butler 1993, 2). The gender identities authorized by motherhood and fatherhood, and reproduced in our children, are achieved mostly through repetitive performances. Butler asserts that all performances are a form of drag, especially the socially mandated displays of patriarchal heterosexuality often required by parenthood. Understanding how subjectivities are constructed and politicized through the process of naming a father or a mother is "to call into question and, perhaps most importantly, to open up [the] term, like the subject, to a reusage or redeployment that previously has not been authorized" (Butler 1992, 15). Understanding parenthood and raising children as an unnatural performance may enable us to unpack the rather thorny relations and roles of gender described in the previous section.

Interest in the issues of performativity, patriarchal power, and personal responsibility leads me to go beyond a narrow, functionalist interpretation of the San Diego study findings. I think that patriarchal power relations are embraced as covert cultural norms, and these norms are translated into how men and women feel about their respective responsibilities (Del Castillo 1993). Alternatively—and I don't necessarily see this as a contradiction given the complex layers of meaning that surround family construction—some men are much more committed than others to giving time to domestic roles and helping out after the birth of a child, even although they do not feel that they have the ultimate responsibility for these activities. Thus, gender roles (at least as revealed by time commitments) change after the birth of a first child, but gender relations (with perceived responsibilities) remain underwritten by a patriarchal discourse. This power discourse is clearly a source of stress for some of the women in the San Diego study. At the level of everyday family life, contradictions between collective responsibilities and individual commitments appear to be normalized, at least covertly, rather than resolved.

As I have noted, postmodern thought about family form begins with the pluralism, disorder, and fragmentation that cannot be predicted by the modern paradigm of universal reason and its attendant theories of rational family norms, such as gender-role theory. Stacey (1990, 17), for example, uses the term "postmodern family" to signal the contested, ambivalent, and undecided character of contemporary gender and kinship arrangements. She notes (18) that today there is no contemporary family order:

> We are living, I believe, through a transitional and contested
> period of family history, a period after the modern family order,
> but before what we cannot foretell. Precisely because it is not
> possible to characterize with a single term the competing sets of
> family cultures that coexist at present, I identify this family
> regime as postmodern. The postmodern family is not a new

model of family life, not the next stage in an orderly progression of family history, but the stage when the belief in a logical
progression of stages breaks down. Rupturing evolutionary
models of family history and incorporating both experimental
and nostalgic elements, "the" postmodern family lurches
forward and backward into an uncertain future.

Dorothy Smith (1993) and Jon Bernardes (1993) note that the nuclear
family is not only an ideological code supported both tacitly and overtly
by academics but also a conceptual and linguistic minefield and a powerful tool of disempowerment. As a consequence, my conjectures at this
point are speculative and fraught with the same problems of generalization found in many contemporary studies. If nothing else, however, I
am trying to break away from old definitions of gender relations and
roles.

But something else bears on a reconceptualization of Butler's gender performativity: the suggestion that space authorizes some performances and hides others. Some feminists point out that if space is
central to subjectivity, then we need to understand how identities are
constituted, at least in part, by the kind of space through which they
imagine themselves (Rose 1993, 1995; Kirby 1996). I argue here that
mothers and fathers often imagine themselves within the fairly narrowly
circumscribed spaces of the patriarchal bargain. Gender performativity
is a useful concept because it embraces the contradictions that constitute our ideological code and the increasingly diverse and complex strategies young families use to cope with everyday life and an environment
that is increasingly less supportive. It is not enough, however, to consider Butler's work as merely a reconceptualization of men's and women's
gender roles and relations in a more complex form. For families, this
delicate performance must necessarily involve the space of child rearing and include children as social actors performing in a society among
other actors.

Children's performances may be among the most hidden in the patriarchal bargain. Barrie Thorne (1987) was one of the first to point
out that in most feminist analyses children function merely as instru-
ments through which the patriarchal bargain reproduces prevailing gender arrangements, which feminists seek to transform. The space of
childhood, like that of fatherhood and motherhood, needs to be

unpacked because it is still considered a universal, natural, and essential category of existence. The next chapter looks at the relations between childhood and parenthood, and then extends Butler's arguments for destabilizing authorized performances by suggesting the centrality of Lefebvre's notion of a "trial by space." Part of this spatial destabilization, I argue, is a critical recognition of the importance of play.

Chapter 5	Play and Justice

Placing Children within the Patriarchal Bargain

There has been a notable absence until recently of attempts to investigate children and childhood in studies of family structure. Childhood emerged as part of the social imaginary with the beginning of industrialization, but the concern now is for the postmodern liquidation of childhood. The contemporary invisibility of children speaks volumes about the ways the patriarchal bargain continues to authorize certain performances of parenthood and childhood while dismissing others. In considering the implications of a patriarchally constructed motherhood, fatherhood, and childhood, I join the discussion of Butler's gender performativity from the preceding chapter with a consideration of Lefebvre's (1991) "trial by space." The ideas surrounding Lefebvre's work are quite abstract, but they bear a close resemblance to Winnicott's description of the place of play as a "transitional space." I use some of the work by Thomas Herman and myself to suggest that Winnicott's grounded and reasonably intuitive ideas about play can be employed to give specific form to some of Lefebvre's abstract notions of how space is produced (Aitken and Herman 1997). Winnicott's ideas are particularly intriguing when commingled with the work of Butler and Lefebvre because transitional spaces may be theorized as the spaces out of which culture and social justice arise.

The Invisibility of Children

There are several reasons for the lack of empirical and theoretical investigations that focus on children in studies of family structure. First, academic studies and collective family values through time and across space seem to dictate that children follow their parents' decisions. Second, and relatedly, information about children in Western societies is usually subsumed under categories that relate to adults or to the institutions that adults control. Third, many perspectives on children in families come from theories of individual (linear) development that typecast children as "not yet adults," and, as such, they are seen as detached from the social life of which they are manifestly a part. Fourth, we have little systematic information on the experience of childhood in any society (see Aitken 1994). The extent to which children have, or had, an active role in families may need some detailed exploration if we are to understand the social construction of childhood.

The invisibility of childhood in contemporary family studies is particularly irksome given the importance of Philippe Ariès's *Centuries of Childhood* (1962) to the emergence of a social, demographic, and psychological history of families. Ariès argued that childhood as we know it emerged only in the early modern period and was closely tied to the appearance of a spatially separate family identity. Looking back at premodern France and England, he described families as closely tied to the community, and households as open to nonrelatives (servants and apprentices) who were engaged in family activities. "This movement of collective life carried along in a single torrent all ages and classes, leaving nobody any time for solitude and privacy" (411). Children were mixed with adults as soon as they were considered capable of doing without their mothers or nannies. Family portraits—depicting children as little adults not only in their dress but also in their physiognomy—suggested to Ariès that the notion of childhood was foreign to premodern society. In addition, the early apprenticeship of children loosened the emotional bond between parents and children.

The bourgeois classes in mid-eighteenth-century France and England began pulling away from these wider networks of sociability, as Ariès termed it, through the separation of workplace from home and the creation of home as an increasingly private world where new forms of intimacy between parents and children could be initiated. The modern family, he argued, evolved as a private retreat from sociability. Ariès pointed out that the "discovery" of childhood prescribed important changes in family and social structures. These changes led to, and were led by, the changes in families' emotional and psychic structures discussed previously, in which children play an important part. Ironically, while Ariès's work defined the modern family through the discovery of childhood, many contemporary theorists suggest that postmodernity is heralding the demise not only of the family but also of childhood (Hengst 1987; Frønes 1994).

David Postman (1982) defines childhood as a protected and segregated realm generated historically by the educational system and the bourgeois family.[1] Day-care systems and education systems were established to embrace and socialize children in preparation for their entrance into the adult world. The amount of time children and youths spent in educational institutions increased dramatically through the twentieth century, and, importantly, this extension of educational control makes unclear when childhood ends and adulthood begins.[2]

The period of a rapidly shrinking and narrow domestic sphere saw the rise of what some feminists call "social childhood" (Glenn 1987): childhood came to be seen as a special and valued period of life, and children were depicted as innocent and dependent. At this time, the notion of childhood seemed to be going through a re-visioning to complement the changing images of motherhood and fatherhood that I discussed in the chapter 3. At the same time that childhood became thought of as a time of innocence, many family functions moved into the public and institutional realm. As the welfare of children became a public concern, the distinction between adulthood and childhood may have begun to lose its edge.

This public concern is illustrated by a new emphasis on children's formal rights (compare Adams et al. 1971; Okin 1989), the weakening of parental or household authority, and the growth of an institutional apparatus requiring individuation. "A room of one's own for a child is not only a private spatial sphere made possible by increased affluence: it is also a private symbolic sphere, underlying the child's position as an individual and a personality. The individualization of childhood pushes the common categories of 'children' and 'parents' more into the background and stresses the intentions and personality of the individual child or parent" (Frønes 1994, 153–54). Frønes goes on to argue that the culture of the "democratic" modern family is characterized by negotiation and the homogenization of symbols through which decision making and social control take place.

For Postman (1982), the homogenization of symbols is primarily through the power and pervasiveness of contemporary media, but the destruction of the childhood realm is also traceable to changes in public education, methods of upbringing, and how families are now formed. Herbert Hengst (1987) argues forcefully that the current "liquidation of childhood" comes about primarily because the care of children is contracted out to the public sphere. He notes that the similarity of employment conditions for childcare and all other productive activities is an example of the homogenization of the life course. Because child rearing is now predominantly in the public sphere, its distinctiveness as a reproductive activity of the private sphere is lost to a regulated professional environment. David Kennedy (1991), for example, details the developmental issues that arise from uniform, institutionalized childcare centers that bear little resemblance to home environments.

Noting this speculation about the disappearance of a distinctive adulthood and childhood, David Oldman (1994) cautions that because adults control productive activities, children may now exist as a hidden "class" to be exploited. At the same time as children are embraced by institutions outside the home, the modern family ceases to be simply an institution for the transmission of a name and an estate, and children become an integral part of the commodified package of capitalism. One new mother in the San Diego study indirectly referred to this commodification in a discussion of Navy family services:

> [Babies] just don't cost as much as you're afraid they're going to. I think that's a piece of advice I'd offer anyone [whose thinking of becoming a parent] because my brother, my little brother, well just recently his wife got pregnant, and he got really scared that he didn't have enough money, and she was going to have to quit her job, and on and on and on. I sat them down and said, "Hey, they don't cost as much as you think they might." I expected [my baby daughter] to cost an arm and a leg. But right after we got pregnant, the Navy family services offers a course

that they call "Budgeting for Baby." And they sit you down, and they make you take a hard look at everything you spend during the week, during the month. Then they make you factor in all the stuff that you need for the kid, which would be, well, you need a car seat to take her home from the hospital. So they "make" you buy a car seat; they make you buy formula if you can't breastfeed. They say, "Okay, now you have to buy this, now you have to buy clothing, now you've to buy" [waves hand in air]. You know, and on and on! They say, "Okay, now this is what you make, this is what you spend, this is what you're going to have to spend on the kid, this is what you can cut out" [laughs]. I learned that the financial concerns of having a child are just an initial thing. After she's there and you have the initial set-up, the rest of it just follows.

Some critical theorists note that the "rest" contributes to an increasing exploitation of infants and children as a commodifiable class. As the excerpt suggests, having a child may be accommodated in much the same way as procuring a new car. More important, the postmodern "liquidation of childhood" is, in effect, a way for capital to hide an exploited class.

Oldman argues that any attempt to restore to childhood its social character must first free it from "familialization," a process that reduces children to a specific role within the parent-child relationship. Familialization carries with it connotations of emotional use-value, socialization, and invisibility within the private sphere. Oldman agrees with Hengst that children and childhood are becoming less contextualized by the private, domestic sphere and more controlled by a public, commodified sphere, and he shares Postman's views on the increasing similarity of the worlds of children and adults. Oldman differs from Postman and Hengst in defining exactly how children are constituted through tension between the public and private spheres. This tension is seemingly abated insofar as childcare and educational institutions are now indistinguishable from other workplaces. And the tension is also lessened because representational homogenization imposes a consumer culture on both adults and children—a culture mediated by, and presented through, the mass media. It is beyond the scope of what I am trying to do here to detail the implications of children and adults getting to know each other's tastes and values through the media, but clearly there are insidious twists to homogenization of this kind.[3]

Given the way he sees childhood being constituted in the post-modern era, Oldman (1994, 58) is convinced that an appropriate cat-egorization of childhood is that of a class situated in a specific mode of production, which he calls "self-capitalization," whose labor is exploited by adults through the organizational assets that they possess and that they impose on the activities of children. Not only is childhood commodified as an integral part of late capitalism, but children are reified as a class whose surplus value (in terms of interests, education needs, and in some cases, labor) is exploitable. Oldman's analysis is from a structural perspective, and, as he admits himself, this is but one cat-egorization of childhood. Nonetheless, his notion of familialization is important because it highlights that there is nothing "natural" about child rearing, and, importantly, if it is constituted as natural, then child-hood becomes merely another commodified category of late capitalism. If we join this discussion with the controversy over motherhood and fatherhood discussed in the previous two chapters, then a paradox arises because these kinds of categories continue in a postmodern era where we seem to be increasingly unwilling to endure the spatial and gender contradictions inherent in day-to-day parenting.

The Spatial Construction of Childhood and Parenthood

Leena Alanen (1994) argues that because most contemporary feminists see childcare as socially necessary work that is differentially and un-equally divided between the sexes, they tend to objectify children as "other" by focusing on activities that are done to children. Conse-quently, children are positioned as passive participants rather than as active social performers. Much feminist thinking thus begins appropri-ately with mothers' oppression but ends with the objectification of chil-dren. Understanding the impact of children as social agents introduces a new and important edge to family and child research in general and feminist scholarship in particular.[4]

Some feminists share the popular view of Susan Cowan and Mary Katzenstein (1988, 25) that the needs and interests of children constitute only a small part of the larger issue of gendered power relations. Perhaps this view is not surprising given the crippling association between womanhood and motherhood that early feminist writing attempted to expunge. Nonetheless, feminism is well suited to address issues surrounding the marginalization of children in society. Too often, children are subsumed within or hidden behind debates over family values, and this long-lasting discrimination against children within studies of families resembles at least one point of departure for academic feminism in the 1970s.

Why, then, are some feminists so reluctant to incorporate fully the marginalization of children in their theorizing? There is, of course, a close connection between the interests of adults and the interests of children. For one thing, family culture is reproduced through children so that the ways in which they are raised have important ramifications for the worlds they will create. As a result of this connection, Cowan and Katzenstein (1988) argue, the issue of what is good for children becomes complex and thus cannot easily be specified within feminist or poststructural arguments. This may be so, but it is not a reason for dismissing children and issues of childhood from debates about how to study families and communities. More important, from a focus on children and childhood much can be learned about difference and justice.

A SPACE FOR PLAY

To understand more fully the spatiality of contemporary families, we need to appreciate the ways in which the concepts of childhood and parenthood are constituted by Lefebvre's "trial by space." This is a difficult idea to grasp but one that bears heavily on the arguments for spatial justice that I am going to introduce briefly here and then revisit at the end of the book. To anticipate my conclusions for a moment, my belief is that through an understanding of how space is produced for, and by, everyone (children, parents, communities, cities, and nations) we get a grasp of how justice is constituted.

Here is how Lefebvre (1991, 416) describes a trial by space:

> It is a space, on a world-wide scale, [where] each idea of "value" acquires or loses its distinctiveness through confrontation with the other values and ideas it encounters there. Moreover—and

> what is more important—groups, classes or factions cannot
> constitute themselves, or recognize one another, as "subjects"
> unless they generate (or produce) a space. Ideas, representa-
> tions or values that do not succeed in making their mark on
> space, and thus generating (or producing) an appropriate
> morphology, will lose all pith and become mere signs, resolve
> themselves into abstract descriptions, or mutate into fantasies.

A trial by space, then, emerges as a legitimizing process that nothing and no one can avoid. Put simply, the myths of childhood and parent-hood that this and the previous two chapters are trying to unravel have gone through a trial by space to achieve their current legitimacy. This trial is marked by repeated performances on the part of children, adults, families, and communities, but there is also an inherent spatiality to these performances.

Although not unproblematic, Winnicott's notion of transitional spaces is compatible with Lefebvre's trial by space because it does not discount the multitude of fields that influence the manipulation and reproduction of shared space.[5] In part Lefebvre theorizes the ways space authorizes certain performances as social imaginaries that are integrally tied to lived experiences. Winnicott highlights the agency of children (and adults) and children's abilities to make use of space in conceptu-alizing identity, place, and difference. Jane Flax (1990, 116) argues that Winnicott's ideas are some of the most important contributions to post-Enlightenment thinking because they de-center reason and logic in fa-vor of "playing with" and "making use of" as the qualities most characteristic of human being. David Sibley (1995) draws on some of Winnicott's ideas to suggest that we can delineate the transgressive boundaries of the self that are hidden by monolithic conceptions of childhood. To date, geography, family, and child-development studies have missed the ways that play, culture, racial identities, and gender formation may be conflated within transitional spaces (Aitken and Herman 1997).

Although Winnicott's notion of transitional space may seem un-duly reductionist, nonetheless it offers the possibility to mess up and make fuzzy attempts to establish the standard dualisms of subject/ob-ject and self/other. This facet alone makes it compatible with Lefebvre's project (Soja 1996). Transitional space "is not inner psychic reality. It is outside the individual, but it is not the external world. . . . Into this play area the child gathers objects or phenomena from external reality and uses these in the services of . . . inner or personal reality" (Winnicott 1971, 51). A third type of reality that simultaneously separates and unites internal and external existence, transitional space represents an "intermediate area of experience which will not be challenged" (13). Winnicott's framework allows the possibility of a flexible manipulation of meanings and relationships. Objects, cultural practices, and self-images may become elements of transitional space and may be altered as an individual adjusts and updates knowledge throughout a lifetime. For Winnicott, the transitional space is a safe place for experimentation and play because it lies beyond the challenge of society's rules and, as such, is a place from where society's rules may be challenged.

Winnicott considers play to be a universal characteristic of being and its effective use to be a requisite for health: play and experimentation are the primary means by which the infinite stimuli and experiences of the world are reconciled into an individual perspective. This perspective is a structure for handling interactions with the external world (which provide information about that realm) while incorporating, maintaining, and protecting an evolving image of self. Discussion of the significance of play in physical, social, and cognitive development has emerged in Katz's work on children's geographies (1991a, 1991b, 1993). Katz (1991a, 15) characterizes play as a "crucial discursive practice of social reproduction." Play, then, is an important part of family discourse. This discourse is carried out between collective society, as represented through cultural norms, family landscapes, and other artifacts, and the individual psyches of the members of that society. In ideal play, children (and adults) experiment with their cultures, sexualities, and environments in a transitional space that is safe from the consequences of the patriarchal bargain. In the same way that performativity theory privileges the ambiguity of gender and sexuality, so Winnicott's notion of transitional space introduces a site wherein performances may be safely enacted. Within transitional space, reality is moldable, and meanings can be formed, broken, and reestablished.

Winnicott's transitional space is especially intriguing because it may be theorized as a space out of which culture and symbolism arise. Within this framework, culture and symbolism are not immutable structures that

necessarily define children, but rather children may contest these structures in the process of creating future cultures and symbolisms. According to this theory, culture is not conceptualized as Freud's external and
coercive "law of the father," which forces the child to separate from
the mother and embrace an abstract and immutable patriarchal culture.
Rather, the child may be able to bring part of her inner self to the traditions and practices of a culture in order to make use of them. In this
account, the agency of the child may shape her cultural practice. Also,
according to this account, the child's ability to choose and utilize transitional objects begins the process of symbolization.[6] For Winnicott
(1971, 102), the capacity to play and the process of symbolization expand "into creative living and into the whole cultural life of man." Culture, like play, is not only something the child can make use of, it is
also a tradition to which she can bring something of her inner self.

WHAT HAPPENED TO "THE LAW OF THE FATHER"?

The process of emancipation suggested by Winnicott may produce a
family of individuals, but it also emphasizes responsibility and commitment. This emphasis leads to paradoxes that cannot be ignored when
considering the family as an institution and as a social organization
(Winnicott 1964, 1965). As I have noted, the family began as an institution closely connected with community and tradition, and it was
integrated in a patriarchal power structure; it then developed into a social unit formed for emotional reasons and characterized, to an increasing extent, by individualization.

In contrast to the patriarchal, authoritarian family, where the parent not only had the "natural right" but acquired the "political right"
to command his children, in a democratic family the authority of parents is a natural right alone, never a political one. This "right" has a
special, temporary nature in that it dissolves once the child becomes
"self-governing" (Elshtain 1990, 57). The problem with focusing on the
process of individualization in modern families lies with trying to establish the limits of "self-governance"—when and where it occurs. This
problem is particularly irksome because, as noted at the beginning of
this chapter, the point of transformation from childhood to adulthood
has become imprecise in recent times.

The ways in which the process of family socialization (and the individualization of its members) has maintained the hegemony of patriarchy in relations between men and women is convincingly argued by feminists (for example, Foord and Gregson 1986), but less is said about the ways it influences the social relationship between parents and children. Jean Bethke Elshtain (1990, 59) states that, for children, there is little doubt that mothers are powerful and authoritative—although perhaps in ways that are not identical to the ways fathers are powerful—but she maintains that this focus is not necessarily important. She suggests that the ideal of parental authority does not presuppose a gendered identity between the mother and father, as suggested by many feminists, because parental authority is special, limited, and particular. Parental authority may be abused in ways more insidious than any other form of authority, but unless it exists, parenting itself is impossible. Family authority is imperative for Elshtain within a democratic, pluralistic order precisely because it is not necessarily homologous with the (patriarchal) principles of civil society. Just because "the law of the father" continues in civil politics, it does not necessarily have to continue in private politics. A direct hierarchical relationship between the private lives of children in families and the public polity would weaken democratic principles. (This notion of destabilizing the hierarchical links between individuals and families, families and communities, and communities and society through a re-visioning of scale relations is intriguing and will be returned to later in the book.)

Children, Elshtain asserts, need particular, intense relations with adults to help them make distinctions and choices as adults. This need does not necessarily negate the importance of the tension between internal drives and external constraints noted by psychologists such as Winnicott, but it does raise the importance of the ability of child and family contexts to oppose and contest societal norms as a marker of what works in democracy. In most social science theories, differences in sexual and racial identities either are not considered at all or are thought to be irrelevant or subsidiary to the "normal" process of child development. But now feminists and social theorists are arguing for a focus on gender and racial formation and the socialization of children within diverse family settings. Elshtain takes this argument a little further by maintaining that respect of difference (between parents and children within families, and between families and society) is crucial for the maintenance of democracy. "The social form best suited to provide children with a trusting, determinate sense of place and ultimately a 'self' is the family. Indeed, it is only through identification with concrete others that the child can later identify with nonfamilial human beings and come to see herself as a member of a wider human community" (1990, 60). Elshtain, then, rejects any ideal of political or family life that absorbs all social relations under a single authority principle such as patriarchy. She feels that the replacement of parents in families would not result in a consensual world of children with equal rights and the status of adults but rather one in which children become clients of institutionally powerful social bureaucrats through processes not unlike those described by Oldman. Unfortunately, Elshtain is not convinced that the family is secure enough to withstand powerful institutional forces because such security presupposes a widely supportive social and spatial infrastructure, which no longer exists.

When children become accepted as fully social persons and not as "others," the social positions and locations of children in contemporary life and the roles that they are allowed or enabled to play in society will need theoretical rethinking. In practice, parental authority today needs to be flexible and open to the recognition that children frequently have something to teach parents: transformations of family structure flow up as well as down.

The Politics of Difference: Rights, Play, and Justice

As we have seen, Winnicott's notion of play is an important contribution to thinking on child development, identity formation, family structure, and justice because it acknowledges the importance of chaos, intimacy, culture, and creativity—all points that relate to arguments for maintaining an openness to the politics of difference. Ideas of difference come together if we are willing to be at play—in the classic sense of joining in a dialogue with ourselves, our children, and others. Understanding difference in this way has considerable implications for re-visioning adult-child relations. Elshtain's (1990) endorsement of a form of family authority that does not mesh perfectly with democratic principles yet remains vital to the sustenance of a pluralistic culture is important because the particular performances of each child's caregivers—cultural, religious, racial, class-based, ethnic, gendered—affect that child's development, identity formation, and, ultimately, the reproduction of the cultural politics of difference. Winnicott, for the most part, avoids applying his concepts to the construction of gender relations and the formation of structures of dominance within the social order. By overlooking these influences on the way children bridge the gap between egocentrism and the external world, Winnicott becomes complicit in the idealization of knowledge systems, which marks much of the literature on child development. Despite this shortcoming, Flax (1993) sees promise in applying the concept of transitional space to an approach to justice that incorporates questions about gender, race, and domination. Winnicott's theory, while perhaps not moving far enough away from the structures of power to be critical of them, does represent one alternative framework for justice that dislodges reason and an alleged unitary truth from its previously central position.

Flax (1993) has undertaken the reworking of Winnicott's theories in order to apply them to a sustained account of identity formation with a focus on political culture. She weaves a feminist concern for the diminution of relations of dominance with a postmodern concern for the play of differences into an approach to justice that hinges on transitional spaces. "Political life is partially constituted through and made necessary by the tensions generated between two recurring human characteristics: our differences and interdependence" (112). Justice exists as a conceptual framework for understanding difference, including the differences between adults and children. Suspicions about claims for a definite and static, mechanistic version of justice based on logic and reason hinge on the fact that it must be essentialist, exclusive, and controlling (culture as the "law of the father"). As an alternative, a discourse of negotiation open to all is a model of justice that seeks to neither marginalize nor prioritize any one point of view: neither that of children nor that of adults.

Traditional versions of justice typically involve either some hierarchical and arbitrary valuation of difference or, less often, some uniform treatment of difference that, while appearing more equitable, disguises the real and ongoing forms of domination that exist in the construction of gender, ethnicity, and class (compare Young 1990a, 1990b). Rather than being a unitary concept grounded in some external truth, justice, as Flax views it, is a process made up of interrelated practices. "Differences must somehow be confronted, accommodated, or harmonized within a whole that strives to achieve the good(s) for all and in which relations of dominance are minimized" (1993, 112). In order for this process not to result in asymmetric dualisms and for difference not to be used as the justification for hierarchies, a mechanism for consensus building must be available. At one level, transitional spaces fulfill this function while at the same time allowing differences to continue to have value within an individual's existence.

The relationship between play and justice is significant when confronting issues of control and the reproduction of a culture of control. It is part of our culture that children are supervised and directed by adults. The control that is exercised over young people's activities through, among other things, the construction of families and the design of space, influences meanings and practice. How spatial trials are played out in contemporary Western society is the subject of the next three chapters of the book; the ideas of Winnicott, Lefebvre, and Butler provide a theoretical fulcrum for that discussion. To this point, the book broaches the construction of families with some recognition of the ways that this social order is constructed in space, particularly through gender relations and performances. The balance of the book focuses on the production of family space with the invention, and re-invention, of neighborhoods and communities as subspaces within the urban social fabric.

| Chapter 6 | Setting the Nuclear Family Apart |

In previous chapters, I sketched the ways that day-to-day gender and family power relations are constituted within the social imaginary of the nuclear family and the patriarchal bargain. I stressed that despite the mythic power of the family, there is no coherent geography of families. My intent here is not to provide any solid spatial theory for family studies, in the same way that I am reticent to single out any one set of coherent family theories for geographic study. Nonetheless, family studies lack systematic attention to changing valences of space and place, and, concomitantly, spatial theorists are remiss in their consideration of changes in families. My perspective on the production of space parallels how I construct the production of identity in previous chapters: although social relations constitute form and manage space, in a very real sense space is more than an end product of these processes; it is itself a process.

I have suggested that it is useful to think of men's and women's gender roles and relations within the family as performances because they influence identity formation. I now suggest that the reproduction of space parallels the reproduction of other forms of identity such as motherhood, fatherhood, and childhood. And so, in the same sense that we think of gender performativity, we can think of the performativity of space. If we assume space is an active social construction rather than a template for our activities, then it too can be thought of in terms of compromise, negotiation, and struggle because it constitutes, and is constituted by, power relations. It follows, then, that to the extent to which a mythic nuclear-family form exercises control over our social imaginaries it also exercises control over urban space. The present chapter weaves a discussion of the power of urban spatial structures around our conversations with San Diego families. My thesis is that the performativity of the nuclear family requires a space that conflates privacy and exclusion with racism and sexism.

Chapter 5's commingling of work by Lefebvre, Butler, and Winnicott focuses attention onto the production of "real" space and the degree to which ideology is inscribed on that space and then acted out and played with. I consider now the evolution of a perceived "naturalness" in certain urban spatial arrangements. I am particularly interested in the ways the nuclear-family myth nestles within a contrived spatial distancing of the public and the private and how this distancing relates to a fine sorting of the social fabric of cities through the contrivance of *sub*urban spaces. I use the term suburban to refer to a residential sphere that is removed from the larger urban social fabric through processes of exclusion.[1] Our conversations with San Diego families suggest that *sub*urban spaces promote behaviors, attitudes, and gendered divisions of labor that are, for the most part, unworkable because they constrain "the space of everyday discourse" (Lefebvre 1991, 25).

The shattering of everyday space entailed in the construction of segmented social spheres arouses fears and anxieties. According to Lefebvre (1991, 25), around 1910 we began losing the "space of common sense, of knowledge (*savoir*), of social practice, of political power . . . as the environment of and channel for communications." Although I do not wish to make extravagant claims for Lefebvre's project, his purported loss of the space of everyday discourse coincides quite nicely with the incipient *sub*urban filling of empty space at the periphery of (and later within) North American cities. *Sub*urban areas, then, may be seen as exclusionary and depoliticized spaces that enervate common-sense knowledge and shut down the communication needed to maintain a valued lived experience. The assumed naturalness of the separation of the public and the private, the nonresidential and the residential, and the urban and the suburban needs questioning from a Lefebvrian perspective because the segmentation of space supports images, and a series of metaphors, that are unable to touch people's daily lives.

With this chapter, I hope to provide some justification for the immiscibility of *sub*urbia and day-to-day living by looking critically at the segmentation of the urban social fabric, the spatial entrapment of women, and the evolution of separate public and private spheres. A consideration of the larger processes of residential segmentation suggests the inherent sexism and racism of past and contemporary *sub*urban processes. In our conversations with San Diego families, for the most part, questions related to what constitutes an appropriate neighborhood in which to raise children evoked anxiety for the safety of children. As an expecting father wrote on one of our mail surveys, "We have just relocated: according to *Newsweek* we have a quiet and safe neighborhood." Often, this fear for children's safety provoked a desire to create exclusive residential spaces.

In an attempt to understand how the nuclear-family myth is embroidered within *sub*urban space, the second half of the chapter focuses on two white, middle-class families' perceptions of urban residential space and the ways they place themselves within this space. Their stories emphasize the unworkability of the nuclear family within a segmented urban space: Trisha's story highlights spatial entrapment in peripheral suburban areas, and Barbara's story suggests a dissolution of the public and the private. Although these stories are complex, they converge, I think, on a common resistance to private patriarchy. I close the chapter by considering ways that private patriarchy is manifest in *sub*urban spaces, with a particular focus on the contemporary "neo-traditionalist" and "new urbanism" movements.

Segmented Residential Spaces

Important qualifications to the public-private antinomy need to be mapped onto the evolving story of *sub*urbia. It may be argued that powerful forces deeply rooted in Western culture persistently draw people who can afford to do so to private home lives. In addition, it can be argued that the "American dream"—which embraces individualism and family values, social homogeneity, aggressive pursuit of goals, the desire to be near to (and to control) nature, freedom to move, and love of newness—conspires with negative attitudes about inner-city life to produce a desire for *suburbia*.

> There are probably fewer communities better than ours for raising children. Four families home-school on our street. The raising of independent-thinking, respectful, and honorable children is a priority—not double-income parents with little time to instill proper values into their children. All of us teach our children to serve others and to value our community and our country. Potatoes, eggs, flour, etc., are borrowed up and down the street—convenient for us all. We are in a rural area so kids hike, fish, catch crawdads, build tree houses themselves, and that is balanced with a healthy work ethic. The home-schoolers sell eggs, mow lawns, collect cans, watch children to learn "responsibility" with wages and more importantly with "relationships."

This young mother's passions (as described in her response to a survey question) relate directly to an ideology and a set of values that prescribe a specific life choice. Some feminists argue that the "American dream" suggested by this choice is a white, middle-class, patriarchal project that seeks to dominate women as well as minorities. If these feminists are right, it is not difficult to posit the exclusionary nature of the suburban ideal: the owner-occupied, single-family home set in a homogeneous social landscape that excludes "undesirable" families.

SUBURBIA AND THE SOCIAL IMAGINARY OF RESIDENTIAL EXCLUSIVITY

Early experiments with exclusive luxury subdivisions reserved for the upper-middle classes and removed from the "terror" of the city date from the mid-nineteenth century. One of the earliest suburbs was Park Village West in London, which was followed by developments such as Gramercy Park in New York City (1831) and Louisburg Square in Boston (1844). Elizabeth Wilson (1991, 45) argues that since its beginnings, the myth of suburban life has embodied nostalgia for the imagined stability and security of a lost rural past.[2] But it was not until the turn of the century, and particularly with the writing and lectures of British utopian idealist Ebenezer Howard, that the idea of suburbia began to foment. I review these ideas here because I believe that they have some bearing on the contemporary *suburban* social imaginary. Howard's *Garden Cities of Tomorrow* (1902) inspired the financing, building, and operation of a new kind of community on the periphery of the old urban cores. His ideas were practical and coincided with criticism of the excesses of urbanization and industrialization by social theorists in Europe such as Ferdinand Tönnies, Max Weber, and Émile Durkeim. Howard was concerned with inner-city squalor and crime, but his solution profoundly influenced the privatization of peripheral, middle-class residential areas.

Howard's Garden City Movement heavily influenced the modern planning professions of many countries (Mumford 1961). In the United States, private developers propagated and practiced Howard's ideas, but not without significant re-visioning. Evan McKenzie (1994, 8) points out that the most important change in the United States was a favoring of private homeownership over Howard's proposal for community ownership of all real property. Howard's hope for "a new civilization based on service to the community and not on self-interest" conflicted with the developing American social imaginary. The aspect of Howard's vision that was readily accepted in American society was the re-creation of a stable, patriarchal, and exclusively white village community (Wilson 1991, 102).[3]

By the beginning of the 1920s, the door was opened for peripheral suburban development on a larger scale than even Howard would have wished. Economies of scale mean bigger profits, and so large corporations built thousands of houses for short-term profit without any thought for long-term social transformation. If any social thinking underscored this development, it was a continuation of the nineteenth-century ideology that separated the workplace from the family. For example, the planned suburban community of Radburn outside of New York City was marketed in 1928 as "a physical and social landscape that would focus on the family" (Marsh 1990, 151). Radburn was innovative in both physical layout and organizational structure, sharing many of the ideals of the Garden City Movement. It housed 2,500 people in three neighborhoods, each built around an elementary school and all clustered around a single high school (McKenzie 1994, 47). Radburn was perhaps most influential in its separation of pedestrians from cars and its exclusive design for families with children. The community was replete with playgrounds, tot lots, and nursery schools, and houses faced the parks so that children could always be kept in view; mothers, of course, were expected to be the ones doing the viewing and supervising. Radburn is important also for the way that it developed restrictive covenants and private government.

Around the same time as the development of Radburn, but on the other side of the continent, the suburban development of Palos Verdes outside of Los Angeles was advertising a different kind of model: "an adult playground which keeps people young and entertains them" (Marsh 1990, 171–73). The developers envisioned an upper-middle-class community providing golf courses, bridle paths, and access to the beach. Children and family ideals were not part of the promotion of this suburban environment, but, as Marsh argues, when residents arrived in Palos Verdes, family-oriented ideals quickly replaced the developers' "hedonistic-playground-for-adults" model. This change occurred under the tutelage of the "women's club," which prescribed family needs such as park space and play schools. Palos Verdes also had racial covenants that kept out specific ethnic groups and deed restrictions that maintained the single-family character of the homes. A home association served as a form of general self-government, and, as at Radburn, it paved the way for the privatization of family life.

Radburn and Palos Verdes underscore a model of family living and the provision of a safe space for raising children that expects women to bear responsibility for reproductive activities. They also underscore a move toward privatization through community governance and racist deed restrictions.

RACIST FEARS AND TERROR TALES

During the post–World War II housing boom in the United States, the building of homogeneous tract houses, the practices of realtors, zoning ordinances, and exclusionary covenants conspired to reinvent Howard's ideas in the predominantly white and middle-class residential areas of the 1950s and 1960s.[4] The Federal Housing Administration actively promoted this kind of segregation on the theory that creating "homogeneous" neighborhoods was a good way to maximize property values. The term "neighborhood" came into common usage only recently in the United States. Zane Miller (1981) argues that the acknowledgment of neighborhoods as significant entities was a response to the industrialization of society and the creation of specialized subspaces to sort out a city by class. The racial discrimination of exclusionary covenants was effectively outlawed by the U.S. Supreme Court in 1948, but, since then, the rise of homeowner associations, common-interest developments, and other forms of private residential governance has shifted emphasis to class discrimination. For example, less affluent families who might be able to afford a house by pooling resources or renting out rooms are often prohibited from buying in common-interest developments. These kinds of life-style restrictions are justified with familiar euphemisms such as "preserving the character/integrity/stability" of the neighborhood (McKenzie 1994, 78).[5]

The notion of neighborhood character, integrity, and stability is too often conflated with class- and race-based fears. Many residential neighborhoods in San Diego are no longer homogenous by race or class, if they ever were, but residents nonetheless feel an attachment to a social imaginary that derives from an earlier era. For example, nostalgia for the perceived safety of the 1950s' and 1960s' neighborhoods within which they grew up causes some white middle-class parents to project images of the dangers their local area might hold for their children, as these excerpts from a new mother and from an expecting father suggest:

> It would be nice if there was a Welcome Wagon. In America, in the old days, when you moved to a new neighborhood, a Welcome Wagon would drive up and have coupons for things and give you something to eat, welcome you to the neighborhood. But that's a thing of the past. This neighborhood isn't too safe. A number of cars have been broken into, and there are a lot of street people who pose a threat. I am always aware of safety. We could use more parks, less houses, and less street people.

> It's not the kind of, you know you wouldn't want—if you had an older child that was able to go out and ride a bike or something, I wouldn't feel safe about having a kid go out and wander around. The biggest fear you have is somebody is going to come and grab 'em.

The fear of child abductions was mentioned by several parents we talked to. For some it was positioned against the perceived safety of their own childhood. Katz (1993) notes that the evolving "child-safety" social imaginary is often focused on "terror tales" of molestation and abuse; these stories are, for the most part, aggravated by the media's exploitation of a few cases where terrible things happen to young children. Parents are understandably upset by these glaring reports of abductions by strangers, but, in reality, most child abuse happens in the home and most abductions are by known relatives. The social imaginary of terror tales names and maps a fear owned by parents of young children. One result of these fears is the implosion of many contemporary families into themselves. "[The neighborhood] is not very safe. I mean it's all right and we're pretty secure here, but we don't spend a lot of time doing stuff outside of the house. Inside the house it's fine, and hanging out in the yard and stuff, but I don't think it is really safe walking around the neighborhood or even walking to the store. When it gets dark out, I don't think we even walk to the store." While this new father feels safe only in his yard, others are happy with their street but are concerned with the spillover of perceived crime and violence from the next block or street. As one new mother describes it, "I really like the neighborhood, but the extended neighborhood I really don't like. There are only certain areas I go to frankly; I don't go south of El Cajon [Boulevard] much."[6]

While the solution for some is to re-vision their residential safety zones and identify with a smaller territorial area, for others the solution lies with community residents' taking responsibility for themselves by simply moving or by taking practical action in situ. One expecting father expressed both these views as he reflected on life in City Heights, a neighborhood south of El Cajon Boulevard which is characterized by the media as culturally diverse with a significant crime and drug problem:

> It's not really safe around here. We don't plan to be staying here
> very long. Maybe buy another house around here and then
> eventually move up to a better area. I know something I'd like
> to see here: we're working towards a neighborhood patrol.
> We're starting one up actually. They are a good deterrent to
> crime because people are out there with their eyes. There are

several laws being passed. [Neighborhood patrols] are really
effective in going out to the johns as well as the prostitutes.
They can impound your car. Then there are loitering laws in
places where there are common[ly] drug deals. [Neighborhood
patrols] can arrest you and take you downtown. These [new
laws] are going to be enforced real soon. I hate—this may sound
crazy—[the fact that] people have too many rights. It is like in
areas like this: drastic areas require drastic measures. I wouldn't
mind house-to-house searches for weapons—illegal automatic
weapons—I don't have a problem with that! We got home from
Yosemite, and somebody drove by the house and shot a gun in
the air right in front here. There was shell casings out in front
of the car. They're kids mostly, mostly high school kids. These
kids live in City Heights, but I see my neighborhood being just
a couple of blocks. [The gang kids] don't live within a couple of
blocks. Some of them come over though.

These excerpts suggest that fears and anxieties drive the creating
of a suburban idyll urban, suburban, and even rural locales. Terror tales
transcend local boundaries: issues perceived at the larger regional scale
of Southern California have implications for raising children in particular locales. The following is excerpted from our second interview with
a mother living in a new condominium in Rancho Penasquitos, a predominantly white, middle-class, peripheral suburban neighborhood located about twenty miles north of downtown San Diego. She and her
husband bought the condominium a few months before their son was
born. Not only was she concerned about her child's future in the neighborhood, but she also was anxious about the changing racial and ethnic structure of the larger metropolitan area:

It's fine at his age, but we think about his future; you know you
have to think about crime overall in San Diego. We—this will
come out wrong—but we live right next door to low-income
housing, and some of the kids aren't [shrugs and voice trails off].
They probably want to steal and that kind of stuff. For his age
it's fine, I just think we'd be concerned overall with crime as
everyone would be! English is almost becoming a second
language, and he might get lost in the shuffle.

The complexity of this mother's concern with local kids "right next door" and the conflation of those concerns with larger temporal and scale issues clearly undercuts any simple notions of naturalized space and hierarchical scale. Although it may be easy to pinpoint a problem with her apparent prejudices, it is equally important to note that the complex space to which she alludes is still beyond the grasp of contemporary planning wisdom and academic theory.

Although crime is often perceived as a regional problem, some parents place blame for crime, violence, and the gang "problem" at the local level and, particularly, on the shoulders of those teachers, parents, and children who have to live the problem everyday. Larger-scale issues of capital disinvestment and regional economic restructuring are rarely mentioned as root causes of community gang problems. As a young mother commented, "Our community would be better for all concerned if we could find a way to control gangs. Our neighborhood would also benefit from parents keeping track of where their children are! I see so many junior high and high school kids roaming around when they should be at school."

Notions of what constitutes a suburban residential sphere sometimes consolidate around themes of urban crime and terror tales that transcend location. Often, the forms of exclusion that seem to persist most are those directed against families that do not conform to the nuclear-family myth. Of course, many other spatial forms of community exist, and I will explore some of these in the next chapter, but the argument that I want to make here is that the spatial design of many residential areas in urban, suburban, and rural Southern California reifies an isolated and exclusionary form of the nuclear family. In the same way that the performance of motherhood, fatherhood, and childhood constructs "others" who do not perform within the purview of the nuclear-family myth, so suburban residential areas are cut off from those others who may be perceived as threatening because they are, as this new father points out, "not usually in this neighborhood."

> [The neighborhood] is not as safe as it could be. The trolley brings in people that would not usually be in this neighborhood. Other than that it seems to be great.

Searching for Family Space

Now I switch from people's attitudes to their everyday lived experience. I use the stories of two young, white, middle-class couples to highlight some of the immiscibilities of spatial performativity and gender performativity. They suggest that the metaphor of suburbia as a stage on which to perform the nuclear-family myth is problematic because, rather than being a passive stage, space too is an actor in the patriarchal bargain.

The notion of space having performativity does much more than merely tie it passively to the actions of people or suggest that people create and manipulate places and that spatial relations are the end product of that manipulation: performativity gives space a power because it can be understood as something that reproduces itself through generational and ideological repetition. The question of how the nuclear family myth is reproduced through suburban space requires a look at the conflation of space and gender identities, and the ways that space requires repetitions that may be inimical to valued family life. I begin thinking about this repetition with a focus on Barbara and Peter and on Russell and Trisha, and their searches for appropriate places to raise a family.

When they married, and before the twins were born, Barbara and Peter moved into an older neighborhood in suburbia adjacent to where Peter was born. Developed in the 1950s, Allied Gardens is now a fairly large community comprising small, three- and four-bedroom, single-family homes lined along the tops of numerous small canyons. The mostly owner-occupied residences are single-story bungalows and ranch-style houses with attached garages. The sidewalks are generally missing on wide, winding roads that often end in cul-de-sacs. On the periphery of the neighborhood, apartment complexes are beginning to replace single-family homes on major roads. The population is largely middle-income Caucasian. Many of the original buyers, now elderly, continue to live in the area, although there is now significant turnover, with young parents like Peter and Barbara gradually replacing the original owners. As Barbara characterized the neighborhood at our first meeting, "Living in a nice area, we walk every night. It's not like we have to drive somewhere to take a walk. There is nothing I need; everything is here, in this area." By the time of our second interview, she was more familiar with her neighbors. "Not only do we have everything so close to us, but our neighbors are great neighbors; a lot of them have lived here for years. It's a quiet street: it's not a street that people come all the way down here unless they live on the street, so the traffic is quiet. The families that have kids all love our little kids. Everyone is just so nice. It really is a nice place. Being on a canyon, when we have the screen open, the breeze comes through, and it's just really pretty."

For Peter and Barbara, having family close by is extremely important to their sense of place as well as to their ability to cope with the twins. Barbara and Peter have created space within generationally contextualized and "familiar" notions of family. Here is how Peter put it during our first interview:

> I grew up four miles away from here, and that's why we were
> looking at houses [here]. I liked this area because the high
> school I went to fed from this neighborhood and mine; so that's
> why we were looking in the neighborhood I grew up in and this
> one because I really liked these areas. My parents still live in
> the house I was born into. I like this neighborhood more
> because it is an older neighborhood, and you can see a cycle
> where there's a lot of young families moving in like ourselves.

Barbara's mother lives in the neighborhood and comes over every day for nine hours to look after the twins while Barbara works. For Barbara and Peter, the close proximity to family provides most of the child-care support that they need:

> Peter and I are both easy-going people, and we've got great
> family, great support. And I know I've had a lot of people ask
> me if I belong to a twins clubs: I really don't have a need to. All
> my family and friends live here and are very close to us, so
> we've got a great support system. We haven't really let them
> [the twins] stop us [from doing] the things we do. We go to a
> restaurant; we take them with us. They're pretty good kids.

In our first interview, before the twins were born and just after Barbara and Peter moved into their new home, Peter noted that the close proximity of his family provided "a reality check" as well as insight into child rearing:

Welcome to parenthood, two instead of one! I am quickly learning that things [get complicated]: since we've bought a house, I've become more responsible; and the fact that we have two kids coming now, I've got to relax a little bit and not be so stressed out—things will work out! For a while there I would get all stressed out and just watch how others have been affected by having kids, my friends and family, my brother and his two kids; they have had hard times, but things always work out. You gotta enjoy life.

Our second interview with Barbara and Peter was about six months after the twins were born, and, at that time, they still seemed fairly relaxed with their new life style. There seemed to be a "naturalness" to their sense of place and identity with the community. Much of this naturalness appeared to be born out of place familiarity and rootedness and out of generational fixivity and repetition.

The story of Trisha and Russell parallels that of Barbara and Peter in that they also were searching for the quintessential nuclear-family environment. They, too, are white and middle-class with a preference for suburban living. In our first set of interviews, Trisha expressed her frustration with the residential environment of Santee, which precipitated the move she and Russell made to Pine Valley, a small, semirural community southeast of Lakeside. The move to Pine Valley before the birth of their first child, Savannah, satisfied their need for homeownership in a safe, residential environment. As Trisha put it during the first interview, "We needed to buy a house. It's very nice and it's quiet [in Pine Valley]. And plus it wasn't close to anything. Santee's getting scary with the trolley coming in. Because I work in La Mesa, where they just put a trolley station, and the crime rate there is out of control. I mean stabbings, cars stolen, in La Mesa! It's not a good thing!"[7]

When asked to characterize the neighborhood within which they lived as it related to the raising of a new infant, like many of our respondents Trisha and Russell mentioned issues that related to safety and security. These attitudes suggest that the spatial performativity of their nuclear family focuses primarily on neighborhood design, privatization, and the exclusivity of residential life. Here is what Russell (first excerpt) and Trisha (second excerpt) had to say about the design of Pine Valley:

It's a good neighborhood to raise children. [There's] safety, [and]
there's a lot of different animals and stuff like that. The schools
around here are supposed to be pretty good. It's a very *quiet*
neighborhood, except during the summertime when everybody's out doing something. It's a *very* quiet neighborhood, and,
uh, everybody really takes a lot of pride in their homes. And
there are a lot of nice homes right down the street [laughs].
[Interviewer: Anything you'd like to change in the neighborhood?] Ya, I think I'd like some more speed bumps in certain
areas, speed bumps and stop signs.

I like that it is quiet. I like the fact that the houses are separated, you know, it's sort of private. There are quite a few parks:
Kids & Thrills—it is like a little amusement park, and it is just
down the way. There's Muir Park and the hiking trails through
Pine Valley with little arrows and all that stuff. The lake you
cannot walk around, or at least you can't ride your bike. I can
take [Savannah] when she's older, and we can go [on our bikes]
out the highway almost to Alpine.[8]

Russell and Trisha made a conscious decision to follow the quintessential geography of the nuclear family, and they are unabashed about that
decision. Trisha quit her paid employment, and Russell took on more
hours of work to pay for the mortgage.

After the birth of Savannah, Trisha's perspective on the neighborhood shifted slightly as she struggled with the isolation:

There's not community things around here. I mean, neighbors
ask neighbors, "I'm going to the store, do you need anything?"
"I'm going to town, do you need anything?" You have to plan
your day. Like my neighbor, she runs her errands all on two days
a week, and she has to plan them and like draw herself a map of
how to do the stuff. I have the neighbors on our lower side: the
neighbors down there are very nice. [But] I don't feel uncomfortable at all: I could have them over on the Fourth of July for
a barbecue, but I would never feel comfortable going over to
their house for a cup of coffee. Anyway, all of the women on
this street, they all work. Everybody in the whole neighborhood
as far down the street and as far up . . . the street as I can
imagine, they're gone all day. Everybody! The houses are all

empty. [Interviewer: Are there any couples with young children?] Em, yeah, but I don't interact with them. . . . Not until the very end of the street is there a couple with two little children, and they're very into their family. [They have] four sets of grandparents who live in town. So, no, I don't interact with them!

Trisha's family context is different from Barbara's in large part because of her spatial isolation: she told us that she had no support network and seemed to feel that the lack of such a network was directly related to their new semirural location. Nonetheless, during our second interview, she reflected on their previous residence in Santee and speculated that it would have been unsafe for her daughter to grow up there:

> But Santee, it was so congested! It is convenient in one way: if you are young and single and stupid and "who cares," everything is right there. You got Blockbuster [Video]; you can get a hamburger on your bike. But then if you start saying, "Well, they're going to bring the trolley in, and my kids are going to be going to school here." Well, I'd rather get out a little ways.

Our sense during the first set of interviews with Trisha and Russell was that they both had an image of the ideal family and the kind of neighborhood space within which it should be sequestered. The decision that Trisha stay at home with the child was made by both of them, although I think Russell was more comfortable with the idea: Trisha was concerned about being "bored crazy" with only herself and a baby "out in the sticks." Both were open with us and admitted that they favored traditional nuclear-family values and living in rural suburbia. They also seemed to be aware of what was "wrong" with those roles but said that they invariably fell into them anyway with, as Trisha put it, "some bewilderment, laughing and a bit miffed along the way."

THE SPATIAL-ENTRAPMENT-OF-WOMEN THESIS

Trisha's context in many ways reflects the spatial-entrapment-of-women thesis, which grew out of several studies of gender differences that analyzed large, randomly selected samples of North American and European urban populations (Hanson and Johnston 1985). These initial studies focused on the journey to work, establishing the empirical regularity that, for the most part, women's work trips are shorter than those of men.[9] From this finding, some feminists concluded that peripheral suburbs were built for a commuting husband. They noted that the relative pressure women felt while constrained to the domestic sphere and fettered to young children was perhaps the most common explanation for the difference in commuting characteristics (Madden 1981; Hanson and Johnston 1985). Women's productive activities were constrained, it was argued, not only by household and childcare responsibilities and by patriarchal relations within the home but also by limited access to urban resources (Madden 1981; Nelson 1986; Figueira-McDonough and Sarri 1987). Studies in the United States, Canada, Sweden, and France demonstrate that family circumstances have a major impact on women's travel (Fagnani 1983, 1993; Hanson and Hanson 1980; Rutherford and Wekerle, 1988). The low density of American suburban sprawl, it was further argued, not only inhibits travel (particularly when the male wage earner has sole access to the family car), but also socializing (Fava 1980).

The notion that single and married mothers have different travel patterns than men and childless women is challenged by England (1993); her work may represent a change in the way we theorize spatial entrapment. England found that women in Columbus, Ohio, who would normally be thought of as being spatially entrapped (married women with young children) actually had longer commutes than women who theoretically should have been the least spatially entrapped (never-married women without children). England concludes that spatial-entrapment theory is historically contingent and describes patterns of the 1970s rather than the 1980s and 1990s. Although England's methodological basis for generalizing the demise of spatial-entrapment theory has been severely criticized (Hanson and Pratt 1994), she does refine our understanding of the complex spatial constraints women and men face in different spatial contexts and locales.

Several other studies put under empirical and theoretical scrutiny speculations about the universal spatial entrapment of women by suggesting significant ethnic and racial differences (McLafferty and Preston 1991; Johnston-Anumonwo 1992; Johnston-Anumonwo, McLafferty, and Preston 1995). This more recent work refocuses attention away from the constraints imposed by women's general isolation toward a fuller appreciation of the constraints imposed by a multiplicity of spatial contexts and gendered roles. I believe that we need to concentrate on all the gendered and spatial performances that contrive women's stories rather than settle for any form of aggregate explanation.

Spatial entrapment is an ideology with psychological, social, and political dimensions. Trisha's spatial entrapment was not caused by her lack of access to a car but by the coupling of an isolated spatial context, the lack of a support system (including, for a time, her husband), and living in a neighborhood of young families that emptied of people every day. Her context cannot be understood without recourse to the social and psychological dimensions of her motherhood as well as its spatial performance. By the time of the second set of interviews, Russell was working in excess of seventy hours a week to maintain their lifestyle. He was worried about job security with a failing California economy, and he was worried about paying the mortgage. Trisha was a full-time homemaker, and she clearly felt not only that she was isolated and cut off from people but also that much of the independence she had developed while in the world of paid work was somehow incompatible with her present tasks. One of our interviewers noted in her field log that she felt Trisha's personal strength and independence made her reluctant to forge connections in the community.

As I mentioned when I introduce their story in chapter 1, Russell and Trisha's nuclear-family ideal began to unravel just before our third set of interviews. Trisha decided to live with her parents in San Francisco for a while, and they sought help in counseling. Reflecting on their lifestyle in Pine Valley, Trisha had this to say:

> Two years ago I felt like we got this house (and it is small), and we're going to sell it, and we're going to get a bigger one and a faster hot rod, and we got caught up in these things and stuff; and [Russell and I] really clashed that way 'cause I was ready to just stay here—I really like it here—and wear the same clothes and be content if I could just stay home; and he was wanting to be upwardly mobile all the time.

Income-tax structures favor homeowners, but, for the most part, only two-income couples have any hope of qualifying for mortgages and meeting payments in most Southern California suburban neighborhoods.

Middle-class couples who try to maintain a nuclear-family life style on one income often encounter the kinds of problems faced by Russell and Trisha. Even scarcer is affordable and suitable space for low-income households, especially with the rising proportion of single-parent, largely mother-supported households.

As more suburban women seek paid employment to maintain the "American dream house," the sociospatial boundaries between women's productive and reproductive work are blurred. Recognition of the breakdown of these boundaries is beginning to inspire studies that emphasize the impact of women's reproductive responsibilities on waged labor (Hanson and Pratt 1995). Broadening the discussion of spatial-entrapment theory in this way involves understanding some of the unique relationships among political, economic, and familial systems. In addition, spatial-entrapment theory considers how constraints are acted out in space rather than how ideology is inscribed in space. If we consider spatiality in the sense of Lefebvre's production and Butler's performativity, then it may be worth reconceptualizing spatial entrapment within particular versions of patriarchy—as described by Lynn Appleton (1995)—and it may also be useful to see how blurred the dualisms of production and reproduction and of public and private become in the everyday lives of families. I return now to Barbara's family context in order to explore the potential dissolution of the public and private for mothers working at home.

HOMEWORK: BREAKING THE BOUNDARIES
OF PUBLIC AND PRIVATE

Some new mothers are able to pursue personal goals if certain child-rearing activities can be supported by a social or contractual environment. Working at home for pay is an increasingly attractive alternative to young mothers and fathers; just over 10 percent of our San Diego sample of mothers and 2 percent of fathers worked out of the home for at least six months after the birth of the new child. Unfortunately, statistics such as these say little about the day-to-day complexities that arise when productive activities are carried out in the private sphere of the home.

Barbara works out of her home every day that she is not traveling for her job. She told us that the company she works for trusts what she is doing and is happy with her working at home, which allows her a great deal of flexibility. Barbara's career is very important to her, and, prior to the twins' birth, she seemed able to prioritize her work and family needs. In an interview during her pregnancy, we get a glimpse of Barbara's organizational skills. "[I] try to keep myself very focused. I block out times where I know I have certain periods of time to make calls, a certain period of time to do that; and once that period is over, I move on. And I prioritize too. I sit and go through all the tasks I have each day and prioritize them, block out time periods for them."

Before the twins were born, she credited her easy-going perspective on this ability to focus, the flexibility of her job, and her capacity to not take things too seriously. "I think everyone gets so hung up. They are so serious that they go overboard. I think that's why I do so well; I'm calm and even-keeled. If something goes wrong you just sit back and look at the pros and cons and try to make a good decision."

Both Barbara and Peter seem to be able to work around each other's schedules quite well. Here is what Peter had to say about their working arrangement before the birth of the twins:

> With my wife out of town and on the road, then I don't have to
> be responsible for mixing what I have to do with what she has
> to do. I didn't have any set schedule yesterday. When she is
> home on the weekends, like today, usually I am up at five in the
> morning so I can go surfing and be home before she is even up.
> Today we have to mix our schedules: what she is going to do
> with when I am going to work in the yard or where we are
> going to go.

For Barbara, the birth of the twins changed many things in her life including her attitudes toward family life and work. Apart from five months' unpaid leave, Barbara's income and time spent at paid employment remained constant after the birth of the twins. She earns more than her husband, although they both work approximately the same amount of time. In spite of what she anticipated prior to the birth Barbara maintains her level of paid employment, including frequent travel. "I haven't let the twins [limit my career] [laughs]. They should, but I haven't let them. I still travel a lot. I'm on the road every other week, out of town, out of state, which makes it kind of hard, but [gives a reflective shrug]."

Her ability to prioritize became compromised with the twins' birth, however:

> [I always have] a lot on my mind. Sometimes I have got too many things going on I can't [becomes reflective and voice trails off]. I have a hard time prioritizing what needs to be done, especially with work. I tend to want to do things at night, and by that time my husband comes home and he wants to relax; and I'm ready to take off [to do work stuff]. And someone's got to watch the kids, and that can be hectic too.

Nevertheless, at six months, the twins fit fairly well into Barbara's schedule:

> They [the twins] usually do whatever I need. If it's something when I'm here at the house, then I'll put one in the playpen, you know, if they're driving me nuts and I need to get something done, or one in the walker, or else put the swing up and put one in the swing. Otherwise I pick one up and carry one with me, or if I go somewhere I take both with me.

On the whole, Barbara likes her balance of homework and looking after the twins, but she notes that much of her calm comes from knowing that they are being well looked after by her mother, who comes to the house every day from 8 A.M. until 5 P.M. "It works really easy. I actually stay very focused with my work, and because I trust my mother, I leave the kids up to her, unless I hear a major catastrophe; then I'll come out."

I have argued that the space of home and work was the creation of exploitative relations between production and reproduction. Of course, the productive/reproductive dichotomy is equally as problematic as the public/private if it is conflated with the opposition of men to women. To defamiliarize these associations, it seems reasonable to argue with some feminists that reproductive activity within the domestic sphere is a societally controlled form of productive activity within which women are fettered and unable to enjoy the political and economic "freedoms" of capitalist production. Although Barbara subverts this thesis by bringing productive activities into the private sphere of her home, the ideological basis of the nuclear family continues to contextualize many of her relations with her husband while an extended-family ethic contextualizes her relations with her mother. Nonetheless, with Barbara, the arguments against notions of "natural" motherhood made in chapter 3 are augmented if we note the connection between her reproductive role and her changing economic and social roles. The complexities of the suburban contexts of Barbara and Trisha suggest a need to revisit the conflation of the suburban/urban and private/public dichotomies.

The assumed naturalness of the public/private, the urban/suburban, and other "related" dichotomies has been described forcibly as problematic by feminist and poststructural geographers (Sayer 1984, 1991; Pickles 1986; Fitzsimmons 1989; Rose 1993; Massey 1995; Soja 1996). Dichotomous thinking is criticized both in general, as a way of conceptualizing the world, and in particular, as a way of dominating women, manipulating sexual politics, and creating racist ideologies. Doreen Massey (1995) argues that dualistic thinking leads to a closing-off of options and to the structuring of the world in terms of either/or and us/them. Gillian Rose (1993) suggests that one form of defamiliarizing the dualisms would be to oscillate between them. As I noted in chapter 4, it may be that fathers and mothers today are oscillating rather quickly between traditional gender roles, but, as Barbara's and Trisha's stories suggest, and as Massey (1995, 493) notes, such "cross-dressing" does not necessarily diminish the power of the dualisms:

> The fact that [dualisms have] been severely criticized in terms simply of [their] descriptive accuracy, most particularly from a feminist perspective, has not destroyed [their] power as a connotational system. What is at issue in the ideological power of these dualisms is not only the material facts to which they (often very imperfectly) relate (many women don't like housework and many female paid employees negotiate a work/home boundary) but the complex connotational systems to which they refer.

One way of coming to terms with the difference between the kinds of space apportioned by dichotomous thinking is through Lefebvre's and Butler's work. If we agree that Western society is characterized by a set of abstract spaces that are fragmented into subspaces devoted to the performance of specialized, homogeneous activities, then Lefebvre and

Butler interpret this process, respectively, in terms of the evolution of modernity and capitalism and in terms of sexual identity. Both would suggest that the processes of fragmentation and specialization/homogeneity need to be opposed, that the boundaries of the dualisms need not only to be blurred but to be overcome.

Soja (1996, 7) points out that one of Lefebvre's most important ideas is a deep critique of binary logic, with his insistence that two terms (and the oppositions and antinomies built around them) are never enough to encompass the practical experience of day-to-day living. In taking up Lefebvre, Massey (1995, 487) makes the important point that these dualisms inscribe not only an abstract space but also a space within which people have to live. "The focus on dualisms as *lived*, as an element of daily practice, is important, . . . for philosophical frameworks do not exist 'only' as theoretical propositions or in the form of the written word. They are both reproduced and, at least potentially, struggled with and rebelled against in the practice of everyday living."

Susan Hanson and Geraldine Pratt (1995, 94) note that the "geographic fission in life that follows gender contours" cedes enormous power to space in contemporary society. Distance, in a real and a metaphoric sense, was the prime instrument used in the development of residential space to isolate women not only from jobs but also from power and involvement in politics. Despite Barbara's example of resistance through homework, the simple thesis of this chapter still stands: suburban implies a controlling distance whereby the objectifier (and controller of space) positions himself dichotomously in relation to the other (woman, child, nurture, emotion, suburbia). The fascination with the other suggests that the "master subject" has an ambivalent relationship to the feminized, racialized and undervalued suburbia. To paraphrase Rose (1993, 77), on the one hand there is fear of the other, of an involvement with the other, which produces a distance and a desire to dominate in order to maintain that distance. On the other hand, there is also a desire for knowledge and intimacy and closeness. In Barbara's example, the spatial distancing is metaphorical, whereas Trisha's spatial entrapment is real and acutely enervating.

Private Patriarchy and the New Urbanism

In different ways, the stories of Trisha and Barbara are constituted within what Appleton (1995) calls a "gender regime." Gender regimes are the product of an ongoing series of struggles within and between social institutions. Irrespective of where they are located, these regimes are usually hostile to changing the patriarchal ideologies of the nineteenth century that created the myth of the nuclear family. As Appleton (1995, 44) puts it, and as the stories of Trisha and Barbara suggest, struggles over gender may be interpersonal and intrapsychic, but they are always characterized by resistance to private patriarchy.

Appleton's (1995) notion of private patriarchy finds its highest form of expression in the suburbs of the 1950s and 1960s, but it is manifest most recently in the rise of "neotraditional" values within the "new urbanism" movement. Within the gender regime of private patriarchy, whether it is located in urban or suburban areas, few alternatives to the monolithic family are acceptable: the hegemony of heterosexuality, parenthood, and marriage is complete, and the main site of gender struggle is the family, as members contests the allocation of scarce resources such as time and money.

One of the foremost external expressions of new urbanism is the exclusion of unwanted elements from residential enclaves through security guards and gated community entrances. As I pointed out earlier, security and child safety are enduring themes of our interviews: excluded elements from supposed child-friendly neighborhoods include automobiles (everybody knows that automobiles and children do not mix), but demands are also made to exclude people who "would not usually be in this neighborhood." These concerns were highlighted when some of our conversations with new parents turned to gated communities:

> Scripps Ranch has everything imaginable (except a theme park) available to young kids. I cannot think of anything that could improve the neighborhood. . . . Okay, . . . make it a completely gated community.

> Only die-hard gated communities appeal to me. Because yesterday the apartment we looked at was in a gated community, and a guy just walked in through the gate. It has to be really gated with a guard for me to feel good.

I don't feel comfortable with cars driving on the street at all. In
fact, we're having a gated community. I'm pretty sure it's going
in. If that happens, I'll be much, much happier.

In Southern California, gated communities may derive from concerns about safety and security, but they are also clearly associated with
neotraditional value systems that place responsibility for child rearing
and commitment to children on individuals, families, and the communities they construct rather than society. When we pressed respondents
who chose to live in gated communities about their relations with their
neighbors, they replied that they rarely talked with them and would
leave their children next door only in an extreme emergency. Developers and marketing agencies sell these communities to new parents
based on their "family orientation" and "old-style" community values,
but those we talked to expressed their feelings in pragmatic rather than
ideological terms. Nonetheless, the conflation of ideology and space produces a fairly rigid gender regime that crystallizes into what McKenzie
(1994) calls "privatopia."

The notion of a new private sphere and an increased segmentation
of exclusive residential areas is evidenced in the increasing number of
homeowner associations and the rise of residential private government.
In the United States, the increase in community-interest developments
(CIDs) and the homeowner associations (HOAs) that govern them is
considerable. Growing from fewer than 500 in 1964 to 130,000 in 1990,
these developments now contain about thirty million people in a variety of different kinds of housing including condominiums, cooperatives,
and retirement communities. By far the most numerous CIDs are
planned-unit developments consisting of single-family homes (McKenzie
1994). Residents own their own houses and lots but pay dues to an HOA
to maintain common areas such as sidewalks, streets, and parking areas. More recently, the ideas behind CIDs and HOAs have combined
with new urbanism to propagate new forms of urban design that incorporate and embrace the spatial and social isolationism with high walls
and security gates. The social imaginary of a nuclear family nested
within these neotraditional citadels is the prominent promotional package circulated by developers and bought into by families.[10] These citadels are an almost insurmountable obstacle for female-headed and
nontraditional households.

McKenzie (1994, 177) notes that CIDs realize part of Howard's social vision in that they proselytize the family, they are antiurban, and, in the United States at least, they have become an alternative form of political and social organization to the extent that they can bring about broad political and social change (at least at the local level). But, "in place of Howard's utopia is privatopia, in which the dominant ideology is privatism; where contract law is the supreme authority; where property rights and property values are the focus of community life; and where homogeneity, exclusiveness, and exclusion are the foundation of social organization." The heterogeneous communities envisaged by Howard for his Garden Cities were different: they were developed to enrich social harmony, promote tolerance, and provide children with a broadened education about the diversity of humankind. Although McKenzie does not broach the preservation of the nuclear-family myth or the spatial entrapment of women and children, his notion of privatopia exemplifies a form of residential development that reinforces both homogeneity and spatial, patriarchal ideologies.[11] Suburban spatial performativity began with a tear in the weave between home and work, but the pervasiveness of privatopias in the contemporary residential landscape constitutes an unraveling of the heterogeneous, urban social fabric.

The public/private dichotomy is linked to an ambivalence embedded in feminism since the nineteenth century and strongly evident today. The ambivalence moves between values of individualism and equality, on the one hand, and nurturance and community on the other. In his vision of privatopia McKenzie does not recognize that the tension between private individualism and public community is as basic to the politics of family change as is the exploitive distancing of residence and work (Thorne 1992, Hanson and Pratt 1995). Feminist geographers criticize the public/private dichotomy because it portrays and perpetuates a false stereotype by ignoring the fact that large numbers of women have always been part of the public, waged sphere. Neither Barbara's homework contestation of the private sphere nor Trisha's spatial isolation in Pine Valley are new stories. For many women, the home has always been a site of production as well as of isolation. Linda McDowell (1983) argued early on that the ideology of separate private and public spheres (particularly a private sphere beyond the reach of capitalism) has led to our missing many important alternative gender regimes.

The debate in contemporary urban design and politics does not accommodate varying forms of identity formation but, rather, implies a reactive vision of self and other, and a consequential conservative view of family and community life. If Ebenezer Howard believed that spatial planning and changing buildings would reform and regularize human beings, so do the promoters of today's new urbanism. Interestingly, many developers of neotraditional urban enclaves selectively appropriate social utopian ideas from Howard and urban aesthetic principles from Jane Jacobs (1961) in order to sell their developments (McCann 1995). The terms of their rhetoric imply an unspoken but powerful belief not only in traditional architecture and neighborhood design but also in the traditional family. Despite unabashed borrowing from Howard, they ignore his commitment to public-land reform and communal living arrangements. Neotraditionalism and the new urbanism produce an equation in which nostalgia for pedestrianism, porches, and white picket fences equals happy families in high-density (but nonetheless private) and traditional (white) communities. These communities are bounded literally by walls and electronic surveillance devices and figuratively by the secure fantasy of the nuclear family.

Like all community forms, they have a power of performance that will arouse emotion and partisanship. Karen Till (1993, 726) sees in these developments "the cultural production of a geography of otherness" because they "reinforce social and spatial divisions by presenting 'differences' as historically, socially, and territorially inevitable." The acceptance of a spatial performance that denies difference invests these suburban spaces with symbolic meaning to the extent that they endure Lefebvre's notion of a trial by space. This trial begs the question not only of how such a space is constructed but of how ideologies of otherness and the private are conflated in the rhetoric of "community."

Chapter 7

Imagined Communities

A ubiquitous rhetoric is found in documents promoting the construction of greenbelt towns in the United States, starting in the 1930s, and the development of neotraditional towns and villages in the 1980s and 1990s. The rhetoric is persuasive, extolling these designed spaces as "safe," "family oriented," "good places to live," and a "return to bonds of authentic community" (Till 1993, 710; McCann 1995). If we are willing to accept Benedict Anderson's (1991, 6) counsel that "communities are to be distinguished, not by their falsity/genuineness, but by the style in which they are imagined," then are we to assume that people are duped by these developers' imaginations? This is a complex question to which answers teeter on the brink of essentialism: there is a certain mystique associated with and a desire for small groups in which individuals live close to one another in local arrangements of personal acquaintance and mutual aid; community space can establish networks of social communication, which is important in the same sense that the intimacy and companionship of living with a few others in the same household is. But recognizing that face-to-face relations in locally based communities are valuable is quite different from proposing them as the basis of a whole society.

Why do people strive for a mythic community ideal? If we assume that "community," with its vague yet generally nurturing meanings, is usually something people desire, then it must be something people do not yet have in the way that they want. In that it is sought after and chosen, community is often imbued with a certain amount of commodity power that cannot be claimed by the notion of "family." For example, some families are attracted to media-perpetuated images of communities where children can be nurtured in safe environments replete with walls, security gates, and speed bumps, while their fear of urban violence is refracted onto the communities portrayed on reality television shows such as *America's Most Wanted* and *COPS*. Neotraditional residential developments advertise immunity from incidents of urban violence in private homes that are built around public space and recreation centers, settings that supposedly foster friendly and caring relations between families. I argue here that the image of inner-city life portrayed by local news media and reality television is complicit with the utopian images of the new urbanism. I suggest with Kevin Robins (1996) that reality television and suburban commodification conspire to corrode "reality" so that our sensibilities are anaesthetized by images that push back moral experiences in favor of thrill and presence.

Of further importance politically is the generally held conviction that communities, like families, can demand sacrifice. Many people are willing to make the sacrifice required to buy into the ideal of a family-oriented community because it fulfills a desire for social wholeness, security, and identity. This is an understandable dream but a dream nonetheless and, as I argue throughout this chapter and the next, one with serious political and geographic consequences. Ironically, at the same time that our society encourages sacrifice, it also discourages and destroys communities of friendship, just as it squeezes and fragments families.

I begin in this chapter to broach more fully the issues of community, difference, and justice. I start with some common wisdom about what constitutes community—ideas that derive from fin-de-siècle writers and contemporary, postmodern social theorists—and tie these ideas to the moral experiences of a Hispanic family in an inner-city San Diego neighborhood, Logan Heights. I then briefly trace some academic debates on place identity and community aesthetics and counter these with a description of Doreen's creation of community in a single-parent context in another urban neighborhood, North Park. The chapter closes with a discussion of academic responsibility for perpetuating a certain image of community.

Some Changing Geographies of Social Organization

Tönnies's ([1887] 1957) distinction between gemeinschaft (community) and gesellschaft (society or association) informs a large part of this century's discussion and debate about what constitutes community. In many ways, this dualism is a modernist foundation for academic thinking about social and spatial relations. Current theoretical accounts of these relations mostly juxtapose and fluctuate between the poles Tönnies established. The work of Tönnies is similar to that of other social theorists in the late nineteenth and early twentieth century, such as Simmel, Durkheim, and Weber, in that each defined a concrete change between the premodern and modern in terms of changes in social relations. Beyond this important connection, the work of these theorists is quite dissimilar. Although the writings of Weber, Simmel, and Durkheim, like all social theory, are at once individual and a product of their time, they need to be reconsidered here for two reasons. First, the work of these theorists provides a foundation for the contemporary debates between individualism and communitarianism, and so at least a brief accounting is required.[1] Second, something more than a brief account is required because their work speaks eloquently to changing spatial and community relations.[2]

Gemeinschaft is conceived as deep, horizontal social relations wherein people remain united in spite of all separating factors. Controls over individual behavior are exerted through the informal discipline of family and neighbors. Tönnies maintained that these forms of social interaction largely colored relations in premodern society. Place is valued as primary, in part, because limited spatial mobility fosters parochialism and intensive, face-to-face, local social relations. Within this circumscribed space, people interact frequently in highly personalized ways. Gesellschaft does not include a real community for contemporary family life as such. People are separated in spite of all uniting factors such as spatial proximity or common needs. Social relations are founded on rationality, efficiency, and the contractual obligations of capitalist organization. In their separateness, individuals find community among common-interest groups rather than through neighboring and local socialization. Any communal arrangements are of limited liability, and most daily, face-to-face contacts are on a contractual basis.

Such limited-liability communities thrive insofar as members share a physical landscape and have similar material goals, such as the maintenance of property values. Individuals follow particular life paths and, in their separateness, join with self-interest communities.

Weber ([1904–5] 1976) traced the processes of gesellschaft by suggesting that Western capitalism located modern productive organizations linked to rational market forces in communities where they had not previously been known. In time, these communities evolved into the classes or interest groups that became the typical basis of social relations in modern society. Weber also accurately predicted the monumental role played by bureaucracies, given the increased complexity of social and spatial relations.[3] Simmel and Durkheim took on the task of characterizing the complex modern modes of social and spatial organization. Simmel's focus was on the money economy, whereas Durkheim emphasized the division of labor. They differ from Weber in that they both identify urban morphology, particularly changes in scale, as an important aspect of spatial organization. Simmel would argue along similar lines to Young (1990b), for example, that the creation of large-scale bureaucracy liberates individuals and groups in ways that support difference.

It may be argued that Simmel's observations on the fleeting, intense, and diverse nature of modern social and spatial relations now shifts up several gears in a postmodern era of mass air travel, modem links, virtual communities, the Internet and the World Wide Web. For example, in words reminiscent of Simmel, Lefebvre (1984) notes that everyday life in late-capitalist society is in a state of psychological terror in part because it is constantly under attack from advertisements, TV commercials, and the apparent need for near-instant communication. These "terrors" include pressure to consume specific products or to communicate in, or look, a particular way. The postmodern self has become a multiplicity of signs anchored in media representations, consumer ethics, and communicative angst (Aitken and Zonn 1994).

According to Robins (1996, 86), the visual technologies that create cyberspace and virtual communities give rise to a gemeinschaft utopianism precisely because "no place" is "every place." He questions the nature of this desire, which constantly seeks the "other" of any place that cannot be satisfied by real place (17). I note, with Robins, that the utopian myth is always viewed as an escape from reality. With postmodern images and communication technologies, however, the myth is given the possibility of a virtual existence. In looking for the origins of this development, Robins's chilling narrative sandwiches our need for community between the development of war technologies and our consumerist needs for reality television. Reality is corroded, he argues, and our sensibilities are anaesthetized by images that push back representations in favor of "presence." In other words, we deny the real and shocking experience of living in a fearsome and violent world by presenting images that do not require reason, analysis, and reflection (121).

For Robins, postmodernism is precisely the tension between disengagement, withdrawal, and solipsism, on the one hand, and the search for belonging and community on the other. Cyberspace's reconciliation of these tensions is dangerous because it is a retreat from reality. Robins highlights the importance of the work of Winnicott with the suggestion that virtual reality creates the illusion that internal and external are one and the same, creating potential space as a third space between inner and outer reality. Although I think it is useful to think of images in the light of Winnicott's work because of the magical-aesthetic aspect of visual technologies, there is a danger in limiting the liberatory potential of transitional spaces that I discussed in chapter 5. Robins is correct to point out that potential space is also a transitional space out of which moral senses and justice may evolve, but this is as far as his arguments dovetail with my earlier discussion of Winnicott. In a limited way, Robins (1996, 95) equates vision with Winnicott's transitional space and thereby concludes that the city no longer functions as a liberatory experience. Our representational systems (including cities and communities), he goes on to argue, are imploded because they remove us from the full experience of reality. I will return to these arguments in the next chapter, but here I examine how Robins's notion of the postmodern presence resonates with those for whom that presence is, in actuality, a moral experience.

MORALITIES OF THE URBAN "SCENE"

In an example that drives much of his theoretical argument, Robins suggests that reality television is a morality play because it publicizes the private lives of its viewers in order to help them deal with the complexities and contingencies of urban life. He points out that television shows like COPS and *America's Most Wanted* seek to connect us visually to the postmodern urban scene, where there is thrill but no morality. "It is a television of the misfortunes, disasters and crimes that afflict urban life. But also a television of how ordinary people cope with and survive them, everyday heroes of the urban scene" (1996, 140–41). For some participants in our San Diego study, that "scene" is a lived, day-to-day experience.

Logan Heights is a working-class, Chicano, inner-city neighborhood of San Diego that has been featured on at least one reality television show. Benito Sanchez lived with his parents, one of his brothers, and his niece Sandra, who was expecting a baby, when we first contacted his household in Logan Heights. This diffuse and relatively unbound extended family varied between four and eight people during the time we knew them. By the time of our second interview, Benito had left for a job in Los Angeles. He was replaced in the household by another uncle and Sandra's boyfriend. The family purchased a small, three-bedroom, detached house just a few months prior to our initial contact. Originally from rural Puerto Rico, the Sanchezes moved to San Diego from New York in 1973 to escape what they perceived as rising violence.

In an interview after the birth of his grandniece, Benito suggested why he thought so many teenagers roamed through their community. He had significant concerns about the deterioration of Logan Heights and gang violence, but his comments suggest that the fear is, at least in part, aggravated by local media:

> This is Logan Heights. Don't you hear the news? Gunshots! I
> mean kids hang out on the corner getting people to buy them
> beer from liquor stores so they can get loaded. That's basically
> what goes on in this area. The Boys Club is basically nonexist-
> ent, and the Community Center is not around; so there's no
> place for these teenage guys to go. There's nothing over here, so
> it's like everybody is hanging out on corners drinking, finding
> something to do. It is easier to find something wrong to do
> than something right to do. And if it was in the neighborhood
> and something fun to do, then they'll go here; but since

everything closes down so early because of the crime rate,
everybody's worried. And if things stayed open a little later, like
the park, and if cops didn't chase people out of the park that
late, and had it had more light, later baseball games or something, and a community center. We don't have a community
center here any more; we have Barrio Logan Station [a youth
community center], but that closes like seven or eight o'clock.
Barrio Station is important to [people] like my little niece since
she is bringing up my grandniece. It is important to her because
she knows people in this area [that go there]. She goes with her
boyfriend. It helps them out.

Rodrigo and his wife, the great-grandparents of the new child, enjoyed their positions as patriarch and matriarch of a complex and mobile
extended family. Rodrigo shared Benito's concern for the deterioration
of the neighborhood, increases in gang violence, and the problems of
teenage mothers. Although unemployed at the time of our interviews,
Rodrigo was active within the local community: he served on both the
elementary school and the middle school board of governors. Perhaps
for this reason he was optimistic that changes can occur with better
community educational facilities:

> I'd like to see Memorial, which is down here on 28th Street,
> become an evening school for parents, for adults, for commu-
> nity courses, say, for instance, for Sandra herself. It'd be nice if
> instead of going to the university, to have courses here. And
> parents like myself that want to advance, or want to speak
> English, would have classes. It would be a benefit to the
> community.

During one interview, Sandra reflected on a myriad of personal
changes that accompanied her pregnancy and the birth of the child.
Of greatest importance is her detachment from local gang activity, although she had not quite left that family to rejoin her "natural" family
because she refused to do whatever was required to make a formal exit
from the gang. When we last talked to Sandra, she was trying to stay
away from her friends who were gang members. She was also trying to
get a job so that she could move to a neighborhood with less crime,
violence, and drugs.

As we saw in the previous chapter, some parents we talked to saw a solution to crime and violence in a nostalgic return to some small-town idyll of community living. For Rodrigo, however, there is nothing idyllic or nurturing about rural living. "We used to live in the country and that made us self-sufficient: we took care of all our business. We do the same thing here. I think I have a neighbor here I haven't talked to in ten years. That [rural] mind-set carried over: you look after yourself and your family." The excerpts from our conversations with Benito and Rodrigo Sanchez suggest a community life that cannot be tied to simple notions of gemeinschaft and gesellschaft, nor do they suggest a disassociation from experience and morality as suggested by Robins's critique of visual culture.

DISASSOCIATED SELVES AND THE "EMPTYING OUT" OF COMMUNITY

Scott Lash and John Urry (1994, 13) suggest that our understanding of the spatial and temporal disembeddedness in the postmodern city stems in part from the work of Émile Durkheim and Marcel Mauss ([1903] 1963). Durkheim ([1893] 1964) hypothesized two forms of social connectedness: mechanical solidarity and organic solidarity. Mechanical solidarity is based on the similarities among people, and organic solidarity, stemming from the division of labor, is based on differences. Durkheim believed that the loss of mechanical solidarity provided by the family and neighborhood is compensated for by the growth in organic solidarity around the workplace. But, with unemployment in neighborhoods like Logan Heights in excess of 70 percent, the power of the workplace does not compensate for the lack of community and family integrity. If Durkheim's hypothesis about the benefits accruing to the organic solidarity of the workplace is flawed for people like Benito and Rodrigo Sanchez, Durkheim's collaboration with Mauss ([1903] 1963) helps us trace a change from embedded community ties in premodern times to a spatial form characterized by disembeddedness, or "emptying out" (Lash and Urry 1994, 14). Durkheim and Mauss's work ties in critically with Lefebvre's (1991, 25) project on the production of space because it suggests that, with modernization, space lost the categories through which people could classify the world. The lost categories comprised the everyday ways that people apprehended "commonplaces" such as town, history, paternity, and traditional morality. Because people are no longer grounded in the material conditions of social life, they have little recourse but to rely on the abstract metaphors and myths that are left.[4] But the myths and metaphors are hollow and floating, characterized by unworkable behaviors and outmoded codes of values.

Given this disembeddedness, what form do ties with family and community take in modern urban life? In urban settings individuals are in contact with many kinds of people in varied settings. Family life becomes fragmented, and social differentiation brings about commitment to divergent life styles, values, and aspirations. This divergence is accommodated at the individual level by people playing distinct roles in different places and in different social interactions. People present different "selves" in different social contexts. For example, Gill Valentine's work (1993; see also Johnston and Valentine 1995) on lesbian time-space strategies demonstrates how the idea of the fragmented self may join with different and complex notions of community so that individuals can maintain multiple identities in different spaces and in one space but at different times: our behaviors are contextualized performances.[5]

The necessity of subversive acts in the face of a dominant culture that defines some groups as different based on bodily characteristics and sexual preferences calls into question the liberatory nature of modern society. Young (1990b, 124) argues that although the objectification and overt domination of certain groups has receded in our own time, "racism, sexism, homophobia, ageism and ableism . . . have not disappeared . . . but have gone underground, dwelling in the everyday habits and cultural meanings of which people are for the most part unaware." The work of Valentine and Young suggests that some political shifts in economy and culture are far below the surface and are not detectable with superficial examination.

Modern Consumerist Aesthetics and Postmodern Communities

Taken together, the writings of the fin-de-siècle theorists may suggest that place-based communities should not exist in contemporary Western society: scale, the consumptive orientation of modernist aesthetics, and other homogenizing factors of urban areas work against social relations that are based on local areas of mutual dependence. Taking a literal translation of gesellschaft, for example, Melvin Webber (1967) argues that, at least in American society, the idea of the community is now indistinguishable from the idea of society. His notion of "places without propinquity" suggests that society comprises various interest groups based on occupational activities, leisure pastimes, and intellectual pursuits rather than place-based affinities: "The striking feature of contemporary urbanization is making it increasingly possible for *men* of all occupations to participate in the national urban life, and, thereby, it is destroying the once-valid dichotomies that distinguished the rural from the urban, the small town from the metropolis, the city from the suburb" (Webber 1963, 29–30; emphasis added).[6] Webber's focus on communication and on how associated technologies contextualize social interaction was quite influential among planners and urbanists in the1960s and 1970s, and it still finds many supporters today. He contends that the essential qualities of urbanness, and hence society, are cultural in character, not territorial, and that these qualities are not necessarily tied to conceptions that see the city as a spatial phenomenon (Webber 1963, 30). The spatial city with high-density concentrations of people and clusters of activity is, for Webber, derived from the communication patterns of individuals and groups.

It is not difficult to argue that this kind of model is flawed because it fails to recognize the spatial construction of the social with its deep-seated ethos of privatism and long-standing tendency toward segregation. Webber denies diversity in many ways and not least because he believes in the "long term prospect . . . for a maze of subcultures within an amazingly diverse society organized upon a broadly shared cultural base" (Webber 1967, 29). Even a cursory reading of Webber reveals that "shared cultural base" to be the liberal democratic and egalitarian belief in the possibility of narrowing income distributions, spreading educational opportunities, and more and more Americans "flooding into the middle class." As I will argue in the next chapter, perspectives such as these justify policies that benefit private interests and allow them to "parade as universal" (Young 1990b, 10). Those interests that are not middle-class or aspiring to be middle-class, with all its entrenched, traditional family values, are stereotyped and marked as deviant and "other," and dramatized on reality television. Urban development since the mid-1970s suggests not only an entrenchment of the injustices that relate to this view but also widening income and education gaps rather than egalitarianism.

In the same way that visual technologies create "a world whose reality has been progressively screened out" (Robins 1996, 13), communication technology enables not only a decentralization of activities but also a decentralization of responsibility for the urban "scene." Paul Knox (1991, 187–88) points out that the planning profession lost a good deal of its moral authority because of its inability to deliver utopian communities, and, as a result it became increasingly aligned with the development and real estate industries, which at least could render a utopian image. This alliance led to a commodification of planning functions, which, in turn, facilitated the privatization and decentralized governance of urban life described in the previous chapter. Joining with this privatism is a consumerist aesthetic that emphasizes pluralistic, organic landscapes aimed at producing a collage of highly differentiated spaces and settings. Noting its capitulation to commodification and consumerist aesthetics while still holding on to some of the overarching notions of rationality and public good, Knox (1991, 188) suggests that planning is now suspended somewhere between modernist and postmodernist positions.

John Jackle and David Wilson (1992, 260) elaborate on the place-based ramifications of the postmodern aesthetic. They note that localities have become places for concentrating "like-activities" and "like-people" in a society of increasing specialization. Peripheral suburban neighborhoods and gentrified downtown communities are contrived for more affluent groups, each of which display characteristic forms of architecture, landscaping, and security. In many ways, this form of conspicuous consumption delineates a celebration of shared values: "Spatial segregation isolates fragments, and makes the unsavory disappear from sight. Within new suburbs, private space has been emphasized over public, and neighbors remain estranged through lack of common experiences, despite superficial appearances of civility that material possessions pretend" (260). In terms of the aesthetics of this postmodern landscape, Jackle and Wilson go on to point out that "community stands as a kind of reference group with which individuals and families see themselves sharing common cause. Therein lies the sense of solidarity: the houses people live in and the things they use provide the vital messages of apparent commonality. As a means of evading experiences that threaten, socially homogeneous neighborhoods provide bulwarks against disorder. Avoidance of the negative anchors the new sense of community" (260).

In sum, privatopia, decentralized governance, and postmodern aesthetics establish a landscape far removed from the egalitarian ideals of Webber. But if we position gemeinschaft as the opposite of this model of postmodernity, then we must take stock of a series of examples that suggest the existence of primary social relations in the form of kith and kinship and of place-based ties. Rather than modernity and capitalism erasing these notions of community, some argue that they continue to exist.

Women and Class-Based Community Aesthetics

With the suggestion at the beginning of this chapter that our contemporary notions of community are somewhat mythic and politically problematic, I do not wish to suggest that there are not aspects of gemeinschaft that are worth pursuing. Since the early 1960s, some scholars have articulated the belief that although the capitalist system, or at least large metropolises as part of the capitalist system, tend to counteract the traditional social ties within smaller communities, these communities still provide an important geographic grounding for many people. Autobiographical and semiautobiographical accounts of working-class neighborhoods, for example, often focus attention on the positive aspects of community that make up day-to-day life. The novels of Robert Roberts (1971, 1976, cited in Dennis and Daniels 1994) and Ralph Glasser (1986) provide vivid accounts of close-knit, working-class, community life. Certainly, the space of this life is contained by gang territories "on the other side of the tracks," and social activities are circumscribed by a juxtaposition of tedious work and unemployment. In addition, if we are to believe these accounts, there is also a greater publicness to daily life and less privacy in working-class areas than in other communities. Nonetheless, the positive aspects of living in these close-knit communities seem to outweigh negative externalities.[7] Alongside these autobiographical works, significant sociological research has probed the existence of gemeinschaft communities in urban areas, and many academics are convinced that they provide an appropriate model for contemporary planning.

The first major anthropological study to highlight community ideals in the twentieth century was Michael Young and Peter Willmott's *Family and Kinship in East London* (1957). The neighborhoods described by Young and Willmott comprise vigorous, local kinship networks that are territorially based. They note a strongly gendered and familial sense of segregation: mother-daughter relationships are the glue that ties these working-class neighborhoods together. Prior to marriage, young women (like their brothers) rebel against societal structures and strive for emancipation, "but when she marries, . . . she returns to the woman's world and to her mother. Marriage divides the sexes into their distinctive roles" (61). This women's network based on strong family ties exists as a sense of community that offsets the doubts and loneliness involved in reproductive work such as childcare. The social networks of men are focused on the public spaces of work, club, and pub. Thomas Jablonsky's (1993) account of the Back-of-the-Yards neighborhood in South Chicago during the first half of this century also highlights the importance of women's networks. Back-of-the-Yards was made infamous by Upton Sinclair's novel *The Jungle* and was also a "laboratory" for one of the many studies of working-class communities conducted by students of Robert Park and Ernest Burgess at the Chicago School of Urban Ecology.[8] Jablonsky focuses on social relations and spatial scale, accounting for home geographies and street landscapes as well as coherence in the community. Although past studies note that labor movements provide important political cohesion for the community, Jablonsky feels that the day-to-day culture of women's lives provides a much more enduring form of community structure. In fact, he gives less emphasis to labor organization and strikes at the Union Stock Yards as an element of community cohesion than he does to women working at home and women with paid employment.

In sum, much of the ethnographic study of working-class areas suggests that community cohesion and social reproduction are situated largely within women's networks. This is clearly an oversimplified explanation: it is difficult to isolate which community-based factors cause the social reproduction of the dominant cultural discoursed imbedded in patriarchy and capitalism. For example, Paul Willis, in his acclaimed *Learning to Labour* (1981), suggests that the resistance and contestation of youth are an intimate part of the process of reproducing capitalist-class relations. Although working-class communities may provide a spatial forum for this reproduction, they should not be isolated from a wide range of global as well as local factors.[9] Willis's study is of the schooling of British working-class males (the "lads") and their preparation for eventual waged labor. The impersonal structures of gesellschaft that organize modern society must be understood, he argues, as being historically and culturally contextualized. His ethnographic methods focus on an exploration of the subtle nuances, forms of behavior, and manners of speech in the everyday lives of the twelve lads in a working-class school who are linked by friendship and nonconformity.

In mapping a Marxist framework onto the everyday cultural terms of his working-class subjects, Willis is concerned with a tension between their knowledge of the system and the implications of their rebellion against it. Willis's lads had remarkable insight into the nature of capitalist exploitation, but, in learning to resist institutional environments, they establish the kinds of attitudes and practices that lock them into their class position. Willis shows also how the oppositional culture created from institutional experiences resonates with other critical locales. The home and community locale, and parental perspectives, show that the culture is generationally reproduced. The very way they talk and learn how to castigate the system ensures their place in a particular sector of low-level waged labor, foreclosing any possibility of upward mobility. The lads denigrating attitudes toward women (as well as West Indians and Asians) perpetuates a patriarchal control of reproduction as well as bigotry, racism, and segmentation in the workforce. Willis argues that this outcome serves the needs of capital for an unskilled labor force and a surplus population of unemployed while at the same time controlling the social reproduction of the system.

VILLAGE IDEALS

Perhaps the greatest influences on community planning in the 1970s and 1980s were the works of Jacobs and of Gans. In her critique of contemporary urban planning, *The Death and Life of Great American Cities*,

Jacobs (1961) extolled the virtues of locally based community ties to a generation of planners who were disenchanted with the ideas of efficiency and optimization that sent freeways through inner-city neighborhoods and created a public-housing fiasco in the United States. Jacobs recognized local communities as important aspects of people's identity and of their satisfaction with day-to-day life. Chains of influential persons, for example, comprise links through whom information and social norms are disseminated to the community at large. These social links require the growth of trust, which is carefully nurtured over time. Her work differed from that of Young and Willmott (1957), who concluded that older mothers preserve community because they provide a network of experience for young married women. Rather, Jacobs's focus was on the functioning of localities around "key people" such as store owners, priests, and local politicians, who provide spatial and temporal continuity for a particular community. If a place loses these critical people, it also loses its vitality as a community. Other key people may include the elderly and those who regularly sit on stoops or hang out of windows. One of Jacobs's main criticisms was of the way modern urban design undermines the ability of residents to observe street activity to the extent that the social control of criminal activity breaks down. In communal areas, people establish interpersonal contacts, which promote natural surveillance and social cohesion. She argued that these "eyes on the street" are more effective for neighborhood security than police or bars on windows.

Drawing on the work of Jacobs, Jackle and Wilson (1992) suggest that three criteria—the social unit, the purpose, and the place—provide a common definition of community. They note that Jacobs grounded her definitions of community in notions of sense of place and a "landscape vigor" that relates to how the place functions as a community (its purpose). Place is primary because it circumscribes highly personalized interactions and social networks, and severed from these established ties, people and places become ineffectual. "Disrupted, a city's social networks do not heal quickly, for people take years to build significant relationships with one another through propinquity" (Jackle and Wilson 1992, 258).

Intense social cohesion was identified in Gans's (1962) celebrated "urban villages." He saw social cohesion as the result of varied cultural

practices based on ethnicity, kinship, neighborhood, occupation, and
life style. Unlike Jacobs, Gans strongly rejected the idea that local social cohesion is in any way diminished by the scale of contemporary
urban life. However, when Gans extended his work to peripheral suburban communities with his study of Levittown (1967), he saw something very different from the inner-city urban villages of Boston. Of
particular interest here are the conflicts of difference that Gans described within this planned, white community and the ways he suggests
they could be resolved (414):

> People have not recognized the diversity of American society,
> and they are unable to accept other lifestyles. Indeed, they
> cannot handle conflict because they cannot accept pluralism.
> Adults are unwilling to tolerate adolescent culture, and vice
> versa. Lower middle class people oppose the ways of the
> working class and upper middle class, and each of these groups
> is hostile to the other two. Perhaps the inability to cope with
> pluralism is greater in Levittown than elsewhere because it is a
> community of young families raising children. Children are
> essentially asocial and unacculturated beings, easily influenced
> by new ideas. As a result, their parents feel an intense need to
> defend family values; to make sure that their children grow up
> according to parental norms and not by those of their play-
> mates from another class. The need to shield the children from
> what are considered harmful influences begins on the block,
> but it is translated into the conflict over the school, the
> definitional struggles within the voluntary associations whose
> programs affect the socialization of children, and, ultimately,
> into political conflicts. Each group wants to put its cultural
> stamp on the organizations and institutions that are the
> community, for otherwise the family and its culture are not safe.
> In a society in which extended families are unimportant and
> the nuclear family cannot provide the full panoply of personnel
> and activities to hold children in the family culture, parents use
> the community institutions for this purpose, and every portion
> of the community therefore becomes a battleground for the
> defense of family values.

The Levittowners is different from his earlier work on urban villages because Gans focuses on children and family values. It is beyond the scope

of what I am doing here to comment in any depth on Gans's allegations that children are asocial and unacculturated. I argue elsewhere that the social and spatial sophistication of children is evident from an early age (Aitken 1994; Aitken and Herman 1997), and some of this thinking appears in chapter 5.

What is important here is Gans's indictment of the attitude of parents toward the raising of children. The historic value of this work lies in the clear indication of the primacy of a monolithic notion of family that varies significantly across economic classes. Importantly, residents of Levittown do not include single parents or homosexual and lesbian couples within the definition of family. They mention race only as a reason for moving to Levittown (37). Gans concludes (415–16) that Levittowners cannot reconcile issues of pluralism because they fail to establish a meaningful relationship between home and community. In particular, they do not allow government to adapt services to reflect diversity among residents. They do not reject government outright, but they channel it into a few limited functions, most of which protect the home against diversity. Although we can take issue with Gans's laying blame squarely on the shoulders of householders, he nonetheless was one of the first academics to raise the dual specters of diversity and scale since the fin-de-siècle theorists. Importantly, he notes the exclusionary nature of the community found in Levittown and how control of local governance is used as "a defense agency, to be taken over by one group to defend itself against others in and out of the community" (416).

The British sociologist Margaret Stacey (1969), was one of the first to capitalize on the potentially universalizing import of Gans's work. Irritated by purely descriptive studies of unique communities, she suggested comparative analyses of "local social systems," wherein residents are related by bonds of kinship, occupation, class, religion, or politics. Where these bonds are missing, she argues, there is no local social system, or it is partial at best. For example, her study of Banbury, England, in the1950s showed how local patterns of mutual aid and community based purely on proximity are disappearing as people became more affluent and no longer rely on each other (1960). Countering the assertions of Webber, Stacey argued that local social systems were viable cultural products.

From the 1960s onward, sociologists and geographers concerned with planning communities were working on ways to make the physical idea of neighborhood coincide with that of community. Ironically, architectural design and urban planning in the 1970s and 1980s moved in favor of creating "communities" and promoting "neighboring" among women at the very time that affluence was undermining neighboring based on mutual aid and women were increasingly being drawn into waged labor (Roberts 1990, 260).

PURPOSEFULLY CREATING COMMUNITIES OF MUTUAL AID

Many of the ideas already discussed in this section come together in Doreen's story, but I suggest that they come together as much because she was a single mother as because of any singular notion of place identity or existent social networks in North Park. When I introduced Doreen in chapter 1, I noted that North Park is a racially and economically mixed neighborhood with a relatively high crime rate compared with the rest of the metropolitan area. Although it has been described as a urban village, it is not a bounded and self-contained ethnic enclave: it has none of the natural, organic characteristics described in Gans's early work, but it provides a space within which single mothers like Doreen seem to thrive. Over several years, Doreen gathered around herself the spatially fixed, social support group that constituted her community. Doreen's plan from the outset was to situate herself in a neighborhood of friends. "I've got my friends who are really close to me to live in the neighborhood: so one friend's over there [points], one friend over there [points], one friend over there [points], and he'll [nods to the back] just be living right back of me. And these also are really helpful 'cause they don't have any children, and they look at my child as their child. It works out."

The ease with which Doreen's network "works out" is a theme she returns to several times during our first interview after Scott was born. She seems to make the community work for her. Doreen's friends are extremely important for emotional support, and she also relies on their proximity and on their functioning together as an extended family. "Yeah, I do, I do [depend on them]. My one friend, who lives over on Fairley [Avenue], which is just four blocks away, she's taken him overnight. Next weekend, my other girlfriend wants to hang out with him on Sunday, and either they'll pick him up or I'll take him over there."

Doreen is adamant that, for the most part, her friends offer to take Scott because they want to rather than out of any sense of duty or friendship:

> They just do it. I have no family in San Diego here at all. In a way they *are* my family. Maybe that's just my way to have a family around me, I don't know [voice trails off]. I mean I've known these women, well one of them I've known for twenty years and the other for ten years and one for like five years. Then I have another friend who lives again in the area who has a husband and a child. Our children were born two weeks apart. And I take him over there, and they play and stuff. So, I'm really lucky. You know, I think about my mom; she had family but they're kind of a weird family. But I'm so lucky because of my friends. They take him *all the time*. And if I'm sick or stressed or if I really need to study [thoughtful pause]. The other day I had this thing I needed to type up and I was brain-dead trying to write something, and my girlfriend, who happens to be a writer, she says, "Come over, I'll help you." Who else could you brainstorm with but a writer [laughs]? So she typed it on her computer, and we were done in an hour; it was great! When I was typing it up, she was watching Scott; so I really do have that support. I'm pretty lucky. I don't think a lot of women have that. There's my best friend down the street. [Interviewer: Is she a mother too?] No, she's not; but she'd come and she'd take him. I have no doubt. And she was there when the child—she was there in the delivery room. She's been there since day one, and she's not a mother; and she's older than me and probably never will be so she's kinda adopted [my kid] as her surrogate son.

Doreen and her husband, Alonso, separated when Scott was a few months old. Alonso (who was born in Spain) lived in the neighborhood for about a year after the separation, and, during that time, he took Scott every Tuesday and Thursday and every other weekend. When we talked to Alonso prior to the separation, it was clear that he had much less of a sense of a caring network. "Neighbors? Unfortunately we don't have a lot of contact with our neighbors; I think that is a very American thing. These relations are friendly, but we just don't interact."

Alonso planned to play an active role in the baby's life. He was less assured than Doreen about asking people in the neighborhood for help; rather, he planned to take on a lot of the childcare responsibilities by being around more and working out of the house. Alonso's employment continued as part-time (twenty-four hours a week) and Doreen's as full-time (forty hours a week) after the birth. Their incomes did not change, with Doreen earning substantially more than Alonso. Alonso perceived his role with the baby to be that of an active participant as well as a care provider:

> I'm home more often now. And if something happened like she needs to go to the doctor or she needs help or whatever, now I'm more available. Before I would try to fight it; I would do excuses. [To make daily life easier, I want to try to be] a model for [my] baby, with love and with good behavior. [I want to] stress the importance of the father as a care provider. Not to leave everything to the mother because it's traditional. To be more androgynous [laughs].

After Scott was born, Alonso took over responsibility for some of the cooking and food preparation, but he was quick to express his intense dislike for these activities. A year after the separation, Alonso moved to Los Angeles, and we were unable to interview him again. Doreen told me that Alonso now looks after Scott every weekend. By the time of the second interview, she had moved to a new house a few blocks away from where she and Alonso had lived.

Much of the distinction between Doreen and Alonso reflects in part an opposition between a "masculine" sense of self and a "feminine" sense of community, which Carol Gilligan (1982) poses in nonessentialist terms as the opposition between two orientations on moral reasoning. The masculine "ethic of rights" emphasizes the separation of selves from neighbors and (real) relationships based on principles such as live and let live. The feminine "ethic of care" emphasizes relatedness among persons and is an ethic of sympathy and affective attention to particular needs rather than of formal measuring according to universal rules. This ethic of care expresses the relatedness of the ideal of community as opposed to the atomistic formalism of liberal individualism (Young 1990a, 306).

Some aspects of Doreen's sense of place are important to note. First, her attachment to her local area preceded the birth of Scott, but, in spite of this previous attachment, Doreen notes a stronger sense of place because of her more frequent walks in the neighborhood with Scott. "I walk more 'cause I take [the baby] out for walks. I'm probably out in the community more since he's been born than before [pause] because I think it's good for children to get fresh air and get walks and get them out of the car [pause]. I certainly wouldn't have ever gone to parks to watch kids play before I had one."

Second, Doreen's commitment to her local area extends beyond wrapping herself around with friends:

> North Park I like 'cause, well, I spent a lot of time before I
> moved to San Diego living in various cities [pause]. I like
> neighborhoods; I like walking even though I don't think
> Southern Californian [cities are] very pedestrian friendly. But I
> really like to walk. I've traveled in Europe a lot; I really like
> that walk, that neighborhood sense. And it makes me feel
> really safe, and it makes me feel really comfortable. I can go to
> the store and they know me. . . . I like [the] fact that . . . I just
> moved three blocks away from where I used to live, so if I take
> my son for a walk I still see the same neighbors. I like that
> community sense; it's a good feeling because a lot [of this]
> country has lost that. . . . It's just a pity.

She notes, in particular, the friendly greetings she gets at neighborhood stores:

> Well, because they know me. I usually go to [the grocery store]
> on 34th Street. When I just had him, the staff saw me when I
> was pregnant, and then they seen me have him. And they'd ask
> me what it was, especially the women, 'cause, you know, it is a
> real bond with women. One time I went there and he was really
> small, and I was kinda holding him, and I didn't have a carrier
> for him, and I was digging in my purse, and the checker lady
> said, "Oh, let me hold him for you." And like that sense, it's
> just a nice bond; it's a good feeling that we don't have too
> much anymore.

The sense of belonging is accentuated when she wants to get out of herself and her own problems. Local store operators offer Doreen friendly and flirtatious chat, a distraction she enjoys:

There's another little market I go to. I stop [on my way to]
school at City College, which again is right in the neighborhood. I would never go to Mesa or Miramar. [Laughs at what
she is about to say.] When I go to State College it'll be like
traveling to Europe for me [more laughter]. I'm kind of unique
and an individual in that area of staying in a community. I
don't think a lot of people share my idea [pause]. And that
place on 31st and Vine Street, . . . every time I go in there it's
like [pause] it's really superficial; they don't know about my
personal problems, if I'm really depressed, my husband, they
don't know any of that. [This last part is said quite fast.] But it
adds to your day to see a friendly face, or the guys, those Arabs
at the little market, flirt with me, and it [pause] it doesn't
bother me; and it's safe, and they don't have malicious intentions. So I don't feel uncomfortable. I just find it very safe and
secure and soothing.

The safe and secure feeling occurs, however, only during daylight.
Doreen would not go out with the baby after 8 P.M., although she has
no problem being out herself:

Yeah, I know [North Park] has a reputation for crime, but I've
never been bothered. I mean, I'm careful; I don't leave things
open; I close my windows when I leave, and I even leave a light
on or music. When I get out of my car, I make sure I have my
keys in my hand; I'm careful; the porch light is on. I've had my
car vandalized, but that happens anywhere. There's police
'copters. East San Diego is right over there; um, they don't
really bother me 'cause that's part of life nowadays.[10]

Doreen also has some concerns about crime and prostitution in
neighborhood parks, although she is defensive about her statements because she clearly wants to create a good impression of the community
as she sees it:

I don't like [the closest park]. Nothing derogatory towards
homeless people, but a lot of homeless people hang out there;
and I'm not—they don't really bother me—but there's a lot of
alcohol there, so there's a lot of broken glass; so I think about
that! And the balls—I don't know if they are playing with
hardballs or softballs—on a little baby's frame could be a big

deal, so I find that kind of dangerous, and that's the only
reasons why I just don't like that park. And I know that there's
a lot of girls working there; well they're not so much working
there anymore, but there's a lot of girls that work. And I don't
care about that either; I don't care what people do [laughs], but,
you know, sometimes with that kind—it breeds a little more—I
don't really want to find some used condom on the road either,
so I just think about those things! And that's my only reason
why I don't take him there. . . . The Trolley Park has got a
playground for Scott, and it's got a coffee shop across the street
for me, so I can grab a cappuccino and go in there and watch
him play. So it works for both of us. It is just a very diverse
group; there's a lot of Spanish-speaking people; there was some
Russian-speaking people one day; and *I like that*. A lot of people
don't like that; they're afraid. But I personally find that very
endearing. I kinda like color. A guy was just killed there, of
course, too, so you've got to think about that; but [people] get
killed everywhere.

Doreen actively works on the creation and maintenance of her community and her embeddedness. Doreen's community coincides with
those described by Gans and Stacey, respectively, because it is in part
born out of her determination to build a sense of place and a social network, but it differs from their conceptualizations because it is also born
in large part out of the necessities of single motherhood. Gemeinschaft
communities may not be an existent reality, but, if in evidence at all,
they are created through the child-rearing exigencies of people like
Doreen.

Mythic Connections between Communities
and Families

There is now a significant critique of the models of community and community change produced by Gans, Stacey, and Jacobs because they derive, at least partially, from the work of the Chicago School of Urban
Ecology, which, in turn, was heavily influenced by Tönnies (compare
Burgess 1926, 1973; Park and Burgess 1967). The Chicago School's biological metaphors, such as invasion and succession, and how these relate to neighborhood instability were derived from studies of Chicago

and then generalized to many American cities. What is often left unstated by these advocates of community is that Chicago in the early
part of this century (and also, arguably, the Boston of Gans and the
New York of Jacobs) experienced relatively unique processes of community change that were the result of a particular combination of cultural, economic, and demographic conditions. These conditions
probably emerged only in limited areas in the country in the late nineteenth century and early twentieth century and were already disappearing when first "discovered" by sociologists at the turn of the century.
Richard Dennis and Stephen Daniels (1994), for example, note the
skepticism that now surrounds work drawn from the Chicago School's
stereotypical descriptions of transitions from preindustrial to industrial
cities. I suggested earlier that industrialization is not primarily responsible for many of the features of the modern family if only because many
of the most significant "new" features date from the period after 1945.
Our notions of "community integrity" may also be postmodern nostalgia for a gemeinschaft that existed only in a few highlighted places.

The implications of academic study and writing on the connections
among families, communities, and society suggest issues of responsibility that are in part traceable to the Chicago School. In particular, the
idea that interaction between the family and the community needs to
be formalized is demonstrated by the influential work of one of the Chicago School's most prolific writers, Ernest Burgess (1926). Downplaying
the significance of the "unintegrated or loosely integrated family" as a
short-lived aspect of contemporaneous social disorganization, Burgess
asserted that the family was a "highly integrated . . . unity of interacting
persons" that "needed to" interface with the community in important
ways. Significantly, in a later publication, Burgess (1973) notes that individuals and communities could not construct and support family life
entirely on their own. Rather, individuals' limited resources of self-
understanding and social skills needed to be supplemented with, and
upgraded by, the knowledge generated through social scientific research.
These ideas were appropriated by the Chicago School, and they prolif-
erated in scholarship and practice. Translated by family experts into
practical techniques, this research was communicated to, and the techniques were implemented by, social agencies such as child-guidance clinics and marriage-counseling centers in the United States for many years.

The implications of Burgess's work returns me to the discussion in chapter 1 of Dorothy Smith's (1993) poststructural critique of family theory and, in particular, her condemnation of the Standard North American Family (SNAF). Smith argues that most academic texts inculcate the SNAF code in some form and thus reproduce its organization. The code, of course, is not limited to texts on family form but is produced and reproduced in writing projects that embrace the community, the city, and society. In documenting the transference of the family and community as ideological codes from an academic to a practical realm, David Cheal (1993, 8) highlights the modernist and instrumental aspects of Burgess's perspective. First, Burgess felt that emancipation was a condition of progress, although such emancipation could exceed society's capacity for social integration. Second, there was a need to identify normal family forms, which are functional and adaptive and, therefore, are well suited to their environment in comparison with so-called abnormal family forms, which exceed the limits of what is possible. Third, the fascination with disorganization as a consequence of modernization, combined with the belief that disorganization was not a permanent or inevitable condition, led to a conviction that society could be transformed through the implementation of grand social designs. Finally, Burgess put his faith in reason, particularly in the capacity of social scientific knowledge and life-changing technologies to restore social equilibrium. The idea of equilibrium, however, validates the perspective that some equitable norm can be imagined and achieved. Such a perspective denies difference among and between communities by suggesting that a normal community form exists and is desirable and that parents' needs are uniform.

What is modernist about the Burgess conception of community space is not necessarily the characteristics of gesellschaft but rather the making of a distinction between gemeinschaft and gesellschaft. This distinction is also evident in the work of Gans and Jacobs, but it finds its most eloquent voice in the commodification of images around neotraditional values and the new urbanism. What remains unstated but clearly implied in this chapter is that gemeinschaft and gesellschaft are as problematic as any dualism. My reading here comes from a suspicion of any kind of dichotomous thinking, but it also comes from a conviction that the notion of the gemeinschaft way of life is a nostalgic throwback to something that did not exist. Importantly, it is a mythic construction, but it is also a large part of how contemporary society has come to view difference, diversity, and justice. To understand the implications of this view, we need to look closely at the ideas of difference and justice put forward by communitarian proponents of the gemeinschaft way of life. Also hinted at but left unconsidered in this chapter are issues of scale. An understanding of spatial scale may be important for making connections among family, community, and contemporary notions of justice. Although postmodern writers articulate clearly the ideas of disembeddedness, alienation, and chaos, the ideas of scale remain to be problematized fully.

Chapter 8   Difference and Justice

The Place and
Scale of Community

British novelist and social critic Raymond Williams (1985, 65–66) points out that unlike other terms for social relations, such as state, nation, or society, the term "community" is almost never used unfavorably. Given its approbatory character, it seems that community is often used metaphorically to establish a sense of political unity. Neil Smith (1993, 105) notes that identities established at other spatial scales, such as home or nation, are easily rolled into struggles over community, making the term one of the most ideologically appropriate metaphors in contemporary public discourse. By extension, suspicions about the power of community-scale politics may derive from suspicions about hidden agendas, skepticism of the moral "high ground," or cynicism concerning claims to particular and static versions of justice. For example, the privatopias described in chapter 6 do not serve difference because their practice of justice is based on claims of impartiality, which allows what is dominant to rule as universal. Here I extend my discussions of dualistic thinking and *sub*urban metaphors in chapter 6 and of how communities are imagined in chapter 7 to a consideration of the metaphorical scale of community. I pick up on my earlier discussions of diversity by exploring ways in which ideas of difference join with feminist concerns for community and for the diminution of relations of dominance. I am particularly interested in problematizing scale in a way that makes clear which values or ideologies are rolled into community political agendas.

If the social construction of scale is illuminated in this way, then perhaps justice can be conceived as part of a conceptual space for understanding difference.

I begin the chapter by discussing the quite complex communitarian literature about a "just form" for public and private social relations. To compensate for some apparent weaknesses in this literature's conceptualization of justice, I pull heavily from Jürgen Habermas's theory of reasoned action and, in particular, Seyla Benhabib's reorientation of that work to suit communitarian ideals. Throughout this characterization of the literature it is apparent that issues of scale are rarely considered central to the just-form debate. To offset this imbalance, I first outline some feminist and postmodern critiques of communitarianism and then attempt to map issues of scale onto the ensuing characterization of justice and difference. Toward the end of the chapter, I elaborate on a developing social theory of scale that may help us understand the political embeddedness of communities (Smith 1992, 1993; Marston 1995). This discussion draws heavily from the work of Judith Garber and Iris Young and is illustrated by examples of how some women in the San Diego study forge community. My assertion here is that community is much more complex than simply being a political form that mediates the rights of the individual and a larger city, state, or national consensus. Similarly, community is not merely a conduit for ideas and values that trickle down from above. This view suggests a static notion of community stuck in a "naturalized" linear scale that assumes an ascendancy of power from the local to the global. Alternatively, we can explore community as a basis for resistance and the slow transformation of power relations wherein scale is discursively forged and manipulated. In this formulation, power relations are simultaneously and continuously negotiated and contested at multiple scales. I use the San Diego study to illustrate how some women "jump scale" and contest the confines of the local by embracing a critical regionalism.[1] An objective of this chapter, then, is to use a critical theory of scale and the practical, day-to-day lives of some women to sketch a social construction of identity, community, difference, and justice.

Communitarianism, Communicative Action, and Justice

Contemporary communitarian positions cover the gamut from notions of self-identity (Taylor 1989) to ideas surrounding collective justice (Selznick 1992). On the whole, communitarians are linked by a general skepticism concerning the benefits attributed to modernity and to places without propinquity. Elizabeth Fraser and Nicola Lacey (1993) characterize the communitarian perspective as follows: first, communitarians are social constructivists in the sense that they believe that community practices are important sites for the construction of self-identity and political value; second, they want to develop substantial conceptions of collective values such as solidarity and reciprocity, and politically link notions of autonomy and community. The first point suggests that communitarian perspectives have much in common with contemporary feminism. I will critique this connection in a moment, but first I want to elaborate on the second point—the ways communitarian notions of collective values serve ideas of justice and difference.

Issues of justice and difference in communitarianism are problematic because adherents often obfuscate the various ways in which the notion of community is used. For example, Lacey (1994) argues that there is a slippage between "pragmatic" and "ideological" notions of community. From a feminist perspective this slippage is problematic because many forms of community historically exclude single women, minorities, or "families" deviant from the nuclear form. While recognizing that the slippage between ideological and pragmatic claims to community is crucial to its discursive political power, we also must recognize that if the slippage is veiled, then it may in fact maintain a conservative and oppressive status quo. Young (1990b, 124) makes precisely this point when she notes that although contemporary society is discursively committed to equality, at the ideological level injustices toward those categorized as "other" are veiled in everyday habits and cultural meanings. At the pragmatic level, justice is veiled because communitarians pay little attention to those who implement solutions and to how solutions are implemented. As a consequence, Lacey (1994) is concerned that many forms of communitarianism are potentially reactive and politically conservative.

For Lacey (1994), the potential for political conservativism in communitarianism is highlighted in three ways. First, communitarians fail to generate an adequate account of how people may gain access to membership in powerful meaning-generating and decision-making communities. Second, communitarians fail to prescribe the forms of membership that can empower participants not only to speak but also to be heard. Third, a similar lack of critical analysis characterizes communitarian thinking on the question of power relations between communities. I will address the third problem through the work of Iris Young in a moment. The first two issues find some resolution for communitarians in Benhabib's reworking of Habermas's theory of reasoned action.

The moral degradation of modernity is highlighted by Habermas's critique of a modern world colonized by the logic of instrumental rationality and communication technologies that deny face-to-face contact. To offset this alienation and disembeddedness, Habermas (1984, 1987b) calls for a paradigm shift from a philosophy of consciousness and self to a philosophy of language and communication as embedded in his theory of reasoned action. The philosophy of consciousness uses models of instrumental and strategic rationality in the drive for self-preservation. For example, a large part of what consitiues urban and regional planning in the United States and Europe is instrumental and strategic because it relates means to ends and techniques to goals without reflection on the justness of the goals themselves. Often, instrumental and strategic rationality is rooted in a self-oriented, subjective goal to dominate and control nature and other people (Aitken and Michel 1995, 23). The philosophy of language uses models of communicative rationality, which raises the validity claims of individuals in the drive for intersubjectivity. Communicative rationality is oriented toward understanding, agreement, uncoerced consensus, and legitimizing the voices of all members of a community.

Unlike many postmodernists, Habermas would not reject modernist rational planning, information technologies, and communities of choice. Instead he would strive to build connections between rational-technical forms of reasoning and those discourses that reflect cultural, racial, gendered, global, and local identities. Thus, community governance is transformed through a process of communicative rationality, which expands the "notion of reason as pure logic and scientific empiricism to encompass all the ways we come to understand and know things and to use that knowledge in acting" (Healey 1992, 150). Any form of knowledge is a product of human wishes, including the will to power and the practices of negotiation and communication. Validity within the practice of communication is based on the speaker's claims of truthfulness, correctness (when compared with social norms), and sincerity. Communicative action is a move by two or more parties to reach an understanding about a particular context; communicative rationality is based on raising the validity claims of each participant in the decision-making process (Miller 1992, 26–27). This focus de-centers the individualistic and self-interested philosophies inherent in liberal democratic perspectives by acknowledging that community politics operate through consensus and negotiation among collective identities. In this way, the Habermasian project significantly extends communitarian thought toward a consideration of justice and an accommodation of difference.

Benhabib (1992, 71) suggests that a Habermasian project of inter-subjectivity and community would focus on how people use language or communicative actions in a discursive way to convey their own personal and different images of space. "The 'I' becomes an 'I' only among a 'we,' in a community of speech and action. Individuation does not precede association; rather it is the kinds of associations which we inhabit that define the kinds of individuals we become." Speech and action not only convey information but also transmit political and moral meaning and, consequently, notions of justice. The problem arises, then, of understanding how justice evolves from communitarian ideals.

Benhabib (1992) extracts from Habermas's work two sets of arguments that link notions of justice with communitarian ideals. First, Habermas assumes that judgments of justice possess a discernible formal structure because they derive from reasoned communicative action, whereas judgments concerning a just form of community are conflated with ideas of "the good life," which has no discernible structure. Second, Habermas maintains that the evolution of judgments of justice is intimately tied to the evolution of self-other relations. Thus, the formation of self-identity and moral judgments are intimately linked, and, of course, self-identity is also linked to a community of speech and action. The issue of justice and the denial of difference, then, comes down to who gets heard within a community of speech and action.

Central to Habermas's communicative theory of society (1984, 1987a, 1989) is the analysis of how individuals or organizations systematically manipulate communications to conceal possible problems and solutions, maneuver consent and trust, and misrepresent facts and expectations. This analysis forms a critical base for understanding how difference can be denied by some contemporary notions of community. Thus, Habermas's theoretical framework has the potential of revealing how individuals as well as power coalitions exclude or restrict community members from being part of a collective social imaginary. According to Habermas, the purpose of communicative action should not be to disclose one's agenda or to impose it on others but to move toward understanding others' positions and to subsequently act collectively to achieve a common goal. Within this perspective, authoritative knowledge bases (such as those grounded in rational-scientific principles or expert community leadership) are not perceived as immutable facts or accepted societal norms, but instead they are culturally imbedded systems that can be disassembled and reassembled through communicative action.

Habermas provides some intellectual in-roads into understanding the power structures that embroider modernity. In addition, he enriches our social and cultural understanding of modernity in such a way that "neither communities of virtue nor contracts of self-interest can be viewed as exhausting the modern project" (Benhabib 1992, 82). Also, the communitarian commitment to constructing collective values and consensus in a Habermasian project resonates with the writings of some feminist commentators. They point out that it connects with issues such as childcare, education, public safety, and environmental concerns, which underscore a feminist agenda against liberal market-based and social welfare–based solutions (Fraser and Lacey 1993). Nonetheless, several aspects of the joining of the Habermasian and communitarian projects are troublesome from feminist and poststructuralist perspectives. First, Lacey (1994) reminds us that despite Benhabib's commitment to the construction of the self within communities of speech and action, the communitarian perspective on self-identity still either derives from a nostalgic vision of the mythic, premodern, small-scale kinship group or entails a fragmented postmodern world where we live our lives across many competing communities. A tension is created between the space of gemeinschaft and gesellschaft that is difficult to reconcile with expressions of self-identity and community. Lacey's concern is that this dualistic space does not enable one to make sense of fragmented experiences, nor does it suggest how such a fragmented world can hold together at the cultural or historical level. Second, and relatedly, a communitarian construction of the self through communicative action suggests processes whereby masculine and feminine identities may be entrenched. Fraser and Lacey (1993) point out that some forms of feminism and communitarianism can perpetuate the patriarchal structures against which they struggle. For example, some radical feminists advocate the creation of communities in which women are in control of all productive and reproductive activities, but not only does this silence men and children, it also perpetuates patriarchy by maintaining the male/female dualism.

A third problem with Benhabib's reworking of the Habermasian project is that it still, ultimately, moves in a direction of compromise and consensus if not common understanding. Enabling contestation is a large part of the contemporary feminist and poststructuralist agenda, but Habermas would maintain that his project is trying to achieve a system of overarching, universal reason, which would deny the possibility of contestation. Similarly, Benhabib does not abandon modern reason but reformulates the Habermasian project in the direction of "interactive universalism," which is contextually sensitive and open to difference and tries to situate the self within the context of concrete communities and gender relations. In this reformulation, Benhabib embraces feminist criticisms of the Enlightenment's conception of a universally disembodied, white, male subject of rationality. Nonetheless, I am still concerned with a seemingly unyielding focus on communication and consensus within communicative rationality. Clearly, in some contexts it is best to engage in dissent and to preserve difference, even differences in understanding and views. Although Habermas would probably agree with this practical and contextual point, he rarely acknowledges the value of preserving dissent and difference (Aitken and Michel 1995, 24). The value of a particular group's views is often highlighted through discord, but Habermas's communicative rationality and Benhabib's interactive universalism seem to ultimately deny opportunities for dissent.

A fourth problem with communitarian concerns for difference and justice is a disregard for the social construction of scale. The writing of Habermas and Benhabib assumes a homogenized space for justice wherein communication and consensus may evolve. Such a conceptualization denies the practical implications of the social and hierarchical construction of scale, which makes dissent and mobility from one scale to another difficult. Not only does this conceptualization conceal the possibility of scale engendering a metaphorical sense of hierarchy that can disable political dissent, but it also disregards a potentially fruitful way of understanding difference.

In the balance of this chapter, I attempt to distance myself from these deficiencies in the communitarian project by switching gears a little. Rather than focusing on the communitarian project through identity and communication, I want to consider how the social and spatial construction of difference and justice relates to community and scale. The arguments I make pivot on the understanding that began to evolve in the last chapter of community as a political metaphor that mediates the desires of families and those of society. Now, I follow Smith (1992, 1993) and Marston (1995) in problematizing spatial scale more fully as an appropriate language for delineating community in an analysis of difference.

Community and Scale: The Situated Subject and the Denial of Difference

To understand how scale is "naturalized" in contemporary society, it is necessary first to revisit the power of metaphor. Because metaphors are used to clarify and make familiar relations and events that are complex and elusive, in tackling the scale of community there is a need to understand how power relations are mediated by the use of metaphors. In social theory and cultural criticism, spatial metaphors are now an important means through which everyday life is understood (Smith and Katz 1993). As Neil Smith (1993, 97) notes, "Not only is the production of space an inherently political process, . . . but the use of spatial metaphors, far from providing just an innocent if evocative imagery, actually taps directly into questions of social power." It thus seems appropriate to use the ideas that are emerging from this literature to focus on the ways that community is linked to ideas of difference and spatial justice.

Smith (1992, 1993) sees geographic scale as the primary means through which spatial differentiation takes place and justice is served. Scale not only establishes boundaries but can engender a metaphorical sense of hierarchy that may empower certain people and institutions and disable political contestation. The suburban landscapes of the last several decades, in conjunction with contemporary neotraditional and self-governance movements, instill exactly the kind of metaphorical hierarchies Smith is describing. Aspirations for territorial control and the establishment of real boundaries are a clear denial of difference. Ideas of justice that allow the pursuit of territorial control are too limiting because, as Michael Sandel (1982, quoted in Young 1990b, 121) notes, "As virtue, justice cannot stand opposed to personal need, feeling and desire, but names the institutional conditions that enable people to meet their needs and express their desires." The creation of a space and a justice based on this kind of logic hinges on an understanding of life that is essentialist, exclusive, and controlling. This is what Gans (1967, 416) was getting at when he called the Levittowners "inner directed" and their local government "a defense agency" set up to protect "against others in and out of the community." Gans was struck by how little this sense of justice among the suburban Levittowners differed from what Alexis de Tocqueville reported in his travels through small-town, middle-class United States in the early part of the nineteenth century. Typically, these traditional versions of justice involve either some hierarchical, scaled, and arbitrary valuation of difference or, less often, some uniform treatment of difference that, while appearing more equitable, disguises the real and ongoing forms of domination that exist by upholding scale as "natural."

COMMUNITY ACTIVISM AND THE LOGIC OF IDENTITY

Marston (1995) suggests that an analysis of the social production of scale needs to remove assumptions of the ascendancy of power and substitute, rather, a set of complex social constructions. In her empirical analysis of American popular movements between 1870 and 1920 Marston points out that through institutions such as voluntary motherhood, scientific domestic management, and female suffrage, "a women's culture, based upon female consciousness, established metaphorical and literal spaces for women that undermined the artificial distinction between the public and private at the same time that it seriously weakened much of the dominant hegemony, both capitalist and patriarchal, that naturalized the existence of the two spheres." Importantly, these spaces, "inscribed and constituted by a separate women's consciousness, *existed at multiple scales*" (emphasis added). Marston points out that if we assume that the body, the house, the community, the city, and the nation-state are socially and historically inscribed, structured, and instituted spatial scales, then we can shed light on the ways in which women moved in significant ways between male-defined public and private life so as to undermine the cultural and political efficacy of the distinctions inherent within this dualism. Marston's point is that rather than creating a separate domestic sphere, women's culture linked the public and the private in quite complex ways that transcended a simple linear notion of geographic scale that might trivialize place-based politics.

To the extent that women's activism and politics are identified with specific locales, feminist and poststructural critiques can raise issues with geographers' fundamental concern with place. These critiques often do not appreciate the findings of feminist geographers that focus on how women contest patriarchal space at the local scale and create space for democratic, diverse communities in the places that they inhabit (Dyck 1990; Marston and Saint-Germaine 1991). Nor do these critiques appreciate the role of women in the production of space at all scales—family, community, society—especially in terms of politics and practice. Clearly, these subject-based and local studies have serious implications for broader societal discourse. Understanding the ways women contest patriarchy at the local level is a longstanding part of feminist critique, but the ways that this contestation can jump scale (Smith 1993) is not entirely clear.

The enduring face of patriarchy in contemporary society suggests that scale is still considered a natural and unproblematic form of spatial and social organization. The inability of many local activist groups to make headway against city, state, and federal jurisdictions and the dismissal of local women's groups' interest in environmental, domestic, and child-rearing issues speak eloquently to the persistence of a patriarchal political culture. Of late, many feminists have voiced concern that local activism tends to mirror women's domestic concerns (child-care, housing safety, the environment) without any obvious impact on larger political and civic cultures. They argue that because these activities amount to "public housekeeping," they are easily dismissed, at the scale of the city, by the same "city fathers" who shrug off their responsibilities for social and local welfare in the first place. The construction of scale in this way suggests that it is an important, though still poorly understood, aspect of the political geography of neighborhoods, cities, and nation-states.

Although she does not elaborate on the importance of the scale of community, Marston (1995) notes that during the time period of her study the "notion of citizenship was an explicit linking of the home and the wider community. Indeed, the prevailing conviction was that an efficient, standardized, and sanitary home would lessen the injurious impacts of urban growth and change on the municipal government. Thus, renovating or reorganizing all the rooms of the home—or building completely new ones—to incorporate the new standards and technologies of domestic science would create a new space, the positive impact of which would resonate, household after household, throughout the community."

Extending this argument, Lacey (1994) suggests that a focus on community has political value because it resonates with the feminist critique of the dichotomization of the public and the private. She feels that recognition of the political relevance of sources of disadvantage and difference heretofore situated in the private is highlighted by feminism. To pick up my arguments in chapter 3, this recognition is clearly important from a feminist perspective that views female bodies as inscribed with meanings inimical to women's full participation as citizens while male bodies are seen as normal and hence, paradoxically, as enabling men to function culturally and politically as disembodied. Marston (1995) points out:

> Space at the scale of the city is the outcome not only of the structuring forces of the economy and politics—which shape the layout of streets, the form and function of buildings, etc.— it is also negotiated and experienced differently by different groups. For late 19th century middle-class women the city was relatively circumscribed: their public movements were

governed by distinctive cultural conventions. For men of
similar class position, the city was very differently delimited.
Urban space was produced differently by and for middle class
men and women through agreed upon cultural norms. In short,
there are a multiplicity and diversity of spaces produced by
macro level structural forces as well as the micro level forces of
human agency.

THE SCALING OF BODIES AND THE CONTRADICTIONS
OF COMMUNITIES

The problem with highlighting and identifying spaces of local activism is that much of what we mean by community identity becomes
conflated with a scale of the body that continues to disenfranchise
women and deny difference (Young 1990b, 125–26, my emphasis):

> An important element of the discourse of modern reason is the
> revival of visual metaphors to describe knowledge. . . . Sight is
> distanced, and conceived as unidirectional. . . . It is a gaze that
> assesses its object according to some *hierarchical standard.* The
> rational subject does not merely observe, passing from one sight
> to another like a tourist. In accordance with the logic of
> identity the scientific subject measures objects according to
> *scales* that reduce the plurality of attributes to unity. Forced to
> line up on calibrations that measure degrees of some general
> attribute, some of the particulars are devalued, defined as
> deviant in relation to the norm.

The scaling of bodies within community and local activism is criticized by Young because she feels it expresses a desire for the fusion of
subjects rather than honoring situated selves. This fusion of subjects one
with one another, in practice, excludes people who are construed to be
different. Thus, for Young, the scaling of bodies reifies racism, sexism,
ableism, and ageism. Community activism is part of this political problematic in contemporary Western cities because those motivated by it
tend to exclude from their political groups persons with whom they do
not identify. The problem with our search for self-identity in community is that it is often based on a desire for unity and wholeness, which,
in turn, generates borders, exclusions, and dichotomies. These borders,
in turn, establish a metaphorical hierarchy wherein the identity of the

individual is merged, for example, with the family, and the family is subsumed within the values of the community.

The desire to merge together in community generates a logic of hierarchical opposition, a separation of the pure from the impure, the inside from the outside. The production of scale is an active process, but the rational logic of community construction seeks to keep borders and boundaries firmly drawn. Young argues (1990a, 305) further that some ideals of community exhibit a totalizing impulse that would include all.[2] This totalizing impulse denies difference in two ways. First, it does not allow a celebration of the difference within and between subjects. Second, in privileging face-to-face relations, it establishes a model of social relations that are not mediated by space and time distancing. The presumption of face-to-face relations is important and needs further elaboration.

Suggesting that the only authentic social relations are in a face-to-face community detemporalizes the process of social change into a static, before-and-after structure much like that proposed by Tönnies. A mythic gemeinschaft with nurturing face-to-face relations is set in opposition to the present gesellschaft of alienation and anomie. The normative model is wholly utopian because it fails to note that spatial and temporal distancing occur not only in gesellschaft but also in face-to-face relations. For example, all conversations and interactions are mediated by gestures and body language that produce forms of proxemics that may be alienating and abusive. Thus, a nurturing image of community denies difference in the sense that it denies the contradictions and ambiguities of social life. Self and community are the products of social relations in profound and often contradictory ways. Complexity and difference need celebration rather than the homogeneity within subjects.

The communitarian model is equally blind to difference between subjects. It is potentially conservative because its basis in social constructivism collapses to radical relativism if all political positions (and all communities) have equal status and no footing for political critique is possible. By extension, hierarchical oppositions are reduced to some equality, and differing scales of interaction are equally valid and powerful. This is clearly an unrealistic assumption. Although the anticommunitarian position poses the fallacy that any move away from objectivism entails a slide into total subjectivism, Lacey (1994) points out that communitarian arguments may not be subtle enough to avoid the force of such a critique. As we shall see in a moment, Young's compromise, although perhaps naive, provides a solution that is inherently geographic because it implies the production of space and the deconstruction of scale as a hierarchical praxis.

COMMUNITIES OF CHOICE, PLACE, AND PURPOSE
Political scientist Judith Garber (1995) also criticizes the romanticization of traditional community forms tied to place, but she is equally skeptical of communities of choice precisely because they are counterpoised with communities of place. It is easy to argue that when community relations are based primarily on family and kin groups and secondarily on local relations with a church, school, and neighborhood web, they are cast in a patriarchal realm. Garber argues further that although social ties within identity groups may be more consistent with diversity and feminist ideals than ties within place-based groups, identity groups also have gender, class, and racial implications. Clearly, this dualistic framework fails to capture the depth, complexity, and extent of people's community relationships, and it misses the feminist potential of democratic places (Garber 1995, 26):

> The flight from defining place-based community is a contradiction, as a prominent theme of deconstructionism, and what most clearly differentiates it from modernist liberal, Marxist, and communitarian theory is that politics cannot proceed when the ineradicability of conflict is denied in favor of majority rule and consensus. Because this is a cautionary note well worth heeding, taking local community as given is particularly unsettling. It appears that as long as women cannot insert themselves into the dominant ideology, community is judged as [a] fundamentally flawed idea rather than one that has worked badly for women and could be rescued if conflicts over community, citizenship, justice and so on were brought to the fore.

Garber goes on to note that much of the feminist critique of communitarianism shares a broad deconstructionist perspective in its attempt to undermine the modernist search for general and universal models and essential qualities of communities. Deconstruction of social and political life involves recognizing contingency, subjectivity, dissonance, and the need to contest norms. Garber's message is quite clear: no universal and unqualified concepts of community are valid, and experiences of community are equally as valid as ideas of community.

Garber notes that much of the feminist deconstruction literature, in its attempt to criticize communities of place, actually endorses communities of choice: traditional universal (patriarchal) referents are set aside for those based on the standpoint of race and ethnicity, sex and sexuality. She argues that these standpoint treatments of community "suffer from a highly constricted conception of the central subject of their inquiry, taking for given what is actually at issue" (26). As Rose (1993), Young (1990a), and Massey (1994, 1995) point out, both universal referents and standpoint referents arise out of modernism and bourgeois culture, and, for that reason, merely reversing their polarity does not constitute a genuine alternative to capitalist patriarchal society. At the same time that they contest the patriarchy of traditional communities, by maintaining the polarities those who advocate communities of choice give up the ability to define community. Valuing local communities, Garber notes, is not widely acknowledged to be a viable option for deconstructionists who would remove as patriarchal communities of place. Nor do deconstructionists attend to how people regard community and participate in a creation of place, which often involves the contestation of spatial and scalar practices.

In many ways, local places succeed in providing meaningful opportunities for women to express relationships with the public sphere. For example, geographer Isabel Dyck (1990) traces the space-time connectivity of suburban women in Vancouver to show that, at a fundamental level, place making is important. She shows how the notion of motherhood and the sets of practices making up mothering work are interpreted and negotiated as women respond to the social and economic structuring of specific places. In particular, women's interpretation and management of mother-child relationships are linked to structured processes of social and economic organization that become concentrated in the specific and unique relations of the communities within which the women Dyck studied lived. Importantly, women are not passive agents in regard to the unique community conditions that contextualize their mothering, but rather they "modify these conditions and negotiate 'good' mothering practices through the recurrent practices of their everyday lives" (479).

Dyck's empirical findings confirm Garber's theoretical suppositions that "woman" as a category is situationally defined, with its meaning and content mutable Dyck 1990, 479–81):

> Space does make a difference to women's lives, not just in the
> form of physical arrangements adding to logistic problems in
> combining paid and domestic labor, but as central to how social
> interaction is constructed and understood. . . . It is attention to
> meaning and the detailed organization of everyday life that
> enables us to disengage "woman" as an abstract category
> ordering life experiences and chances, and place it firmly as a
> situationally defined identity, an identity, however, which is
> also entrenched in the conservative institution of the fam-
> ily. . . . The unfocused and focused locales emerging from
> institutional activity [such as community-based parenting
> classes and parent-teacher associations] and consequential to
> the daily round of activity are sites of communication, a
> communication which not only relays information, but
> negotiates and shares meaning of available concepts [for
> example, the focused locale of the parenting class where
> "expert" knowledge is presented]. Critical to this negotiation
> and sharing of meaning is morality, as moral rules of doing
> things both constrain and, in turn, are open to new meanings of
> morality. Locales provide the mediating links between identity
> and practice, with the moral dimension of practice operating as
> a means of validating identity.

Garber (1995) would concur with Dyck's nondismissal of locale and community, but she expands these ideas by posing communities of purpose, where shared situations foster local political action. Because face-to-face relations are part of urban life, and, at least at the local level, the cloak of anonymity is difficult to maintain, Garber (1995) proposes a "truce" between claustrophobically local communities of place and the elusiveness of achieving invisibility through group identification in communities of choice.

For women (and men and children), communities are not simply a matter of experience or wishful thinking, but rather they must always be seen as political and cultural forms that define people's feelings about local life. Clearly, such a statement is situated in an ongoing ideological conflict over the meaning of place and the politics of spatial hierarchy. Garber's point is that feminists should not throw out the baby with the bathwater by dismissing the importance of the local scale. At the day-to-day level, local places provide a meaningful basis through which women who may not otherwise understand their lives as having a political dimension can express the complex relationships that tie them to the public sphere.[3]

JUMPING SCALE, CRITICAL REGIONALISMS,
AND THE POLITICS OF DIFFERENCE

Will and Carol lived in a community that they really liked. When we first contacted them, they were expecting twins. Both enjoyed the open, "outdoors," and rural atmosphere of Lakeside. Lakeside is located about twenty miles east of downtown San Diego in an area that contrives a sense of place loosely based on "ranching and rednecks." Will was involved with local horse and rodeo shows. When we interviewed them prior to the birth of their twins, Will was somewhat concerned about his job with the Navy. He had accepted a Temporary Assignment Duty on a shore station so that he could be with his wife during the last few months of her pregnancy, but he was unhappy with his current duties. After the birth, he was to be reassigned to his ship in Long Beach. Will expected to spend weekdays in Long Beach and then commute back to Lakeside to be with his family on weekends. Carol summed up some of their fears during the first set of interviews:

> With military cutbacks, everyone is on edge to see what's closing next. I'm more new to the military than Will, so I'm new to the problem. When Will goes up to Long Beach, we know it is coming [so we can plan] as to whether we'd all go, or I'd stay here and he'd go. After Long Beach, we're a lot at the mercy of what they do with us. My father said never to date sailors [laughs]. [The thought of having] the twins is exciting and overwhelming. Having two at the same age at the same time. I guess we'll find out. It'll hit when we bring them home.

Six months after the twins were born, Carol expressed on the mail-in questionnaire her frustration with lack of sleep and having no free time. She worked up until the day of delivery and then was back to work within six weeks. She took a new position as manager of a small, private elementary school at which she had worked; the school has 129 children in kindergarten through grade 6. Carol works between forty and fifty hours every week. Will was surprised at how little time he now got to spend with his wife, noting that the bulk of his weekends at home from Long Beach were spent helping around the house and taking care of the kids.

We were unable to arrange a second set of interviews until the twins were ten months old. As a student and I arrived at their apartment complex for those interviews, Will stormed out of the door saying he had no time to talk. We learned from Carol that he was filing for divorce. The rest of this story focuses on Carol and her attempt to create community. It is uniquely scalar, starting right at the front door of her apartment and continuing on to the Heartland Mothers of Twins Group, which provides for her a regional community. Her story is similar to that of other mothers who are forced by "special circumstances" to imaginatively redefine and re-create family and community.

Carol likes the area she lives in because she feels that it is rural and child-oriented. She believes that everyone looks out for each other. At the immediate, local scale Carol feels safe because the apartment security guard is a close friend of her parents. All the young families are located in one area of the complex, but she rarely interacts with them because her job takes her away during the day:

> This is kid row—all the landscaping looks awful; the rest of the
> complex looks nice. [We are] where all the noise and commo-
> tion is. We've got some a little older and some a little younger
> [than mine]. The one right across is about three months older,
> and then we've got a three-month- and a five-month-old. I
> interact with the other parents very little; most of them are
> home during the day. [On weekends, I get together with]
> Wendy across the way, sometimes. As far as other mothers, I
> mostly hang out with Mothers of Twins. That has been a
> lifesaver for me. It's the Heartland Mothers of Twins Group at
> Grossmont Hospital. I have a sister-in-law who is a twin, and

her mother gave me the number. It has been around for a long time. The group is thirty-five years old. The general meetings are once a month, and they have a planning meeting that I don't usually attend, and they have a Saturday social once a month. They've done picnics. We went to the roller-skating arena, and everyone took their strollers and roller skated [laughs]; and that was wild. Some of us have a hard enough time on roller skates without having to push a stroller at the same time. But the kids loved it.

Between the local scale of the apartment complex and the regional scale of Heartland Mothers of Twins, Carol has established a support network for caring for her boys when she is at work. Childcare during weekdays is provided primarily by her grandparents, all four of whom live close by.

Based on my salary, there's no way I could afford childcare. When I think about the kind of day my boys are having, it is different [because] they are spending time away from me, but they are also spending time with an extended family. I'm sure they get as much out of it as me. I love it! There's very few children who get to grow up and spend whole days with their great-grandpas and -grandmas.

Carol's story is similar to that of other single mothers in our study in that she had to define and create a community that supported herself and her children. By highlighting a feminist critique of the logic of identity and community activism earlier, I wanted to point out that any dismissal of scale opens the potential for an exclusionary politics of difference. An important point that the work of Dyck (1990) and Garber (1995) raises relates to how women like Carol can jump scales so as to contest the production their own identities.

Several mothers in our study mentioned their involvement with La Leche League, an international organization that provides information and support to breastfeeding mothers. It also brings women of toddlers together in support groups to discuss issues such as weaning. A member of one of the La Leche League groups in San Diego whom we talked to found breastfeeding her son during most of his first year to be exhausting physically and emotionally. "He is not sleeping through the night. I haven't slept for more than two hours at a time since Andy was born. I am even more physically tied and drained by Andy [than at the time of the previous interview]. . . . My sister-in-law and mother-in-law are not very supportive because they think it is weird that I am still breastfeeding." She spoke of getting some relief from La Leche League, but her need for interaction with other people exceeded what this organization could supply. "I feel isolated from the community, from my friends and from many others. My time with my husband is very limited also. When he is home, it is much better for me because I can share the responsibility of the [baby] with him. But we don't seem to get much quality time together. I really miss conversations with adults. Something as simple as that!"

Las Madres is a popular organization for parents with newborns and toddlers that is sponsored by a regional HMO. Its prime purpose is to bring together parents and their children in playgroups to offer "friendship, moral support and a wide array of activities" (*Las Madres* magazine, May 1994, 20). Although sponsored by a regional health organization, each playgroup is locally based and makes its own decisions about where, when, and how often they meet. "Las Madres playgroups are organized by geographic location and the children's birth year. Specialty groups can be organized for single-parents, part- or full-time working mothers, fathers, or parents who prefer to meet on weekends (flexibility is a plus)" (*Las Madres* magazine, May 1994, 20).

Several of the families in our study were involved to some extent with Las Madres, and many others mentioned that they had heard of it. Cindy lives with her husband and baby in a planned, gated community in suburban Rancho Cabrillo. She organized a Las Madres playgroup in 1993 that continues to meet weekly. She insists that her Las Madres group is primarily for "stay-at-home" moms. Cindy and her husband "are thankful" that they were able to "buy into, after a lot of research," a community that "was planned to reflect people and families who both work in paid labor and home labor; it also reflects the incomes of those who can afford such amenities." Cindy refers to what she calls a "safe and active" community where amenities include "a swim and racquet club with free childcare, a baby gym, and family bike paths." She sees the Las Madres playgroup as part of her full-time job as a mom. Cindy takes very seriously her time with her daughter. "[Las Madres] is great for stay-at-home moms. It gets them out of the house and relating with others in the same situation. It was very hard for me to leave my job as a teacher, but for now it was [the] right decision to make. My child is the most important thing in my life."

Cindy's gated community is private and, for the most part, excludes minorities and single mothers. As I pointed out earlier, however, community need not be constituted as an oppressive or exclusive private space—a construction that sanctions an image of passive, familylike, communal cohesiveness. Nonetheless, taken together, the examples of Cindy and Carol suggest that communities that satisfy the conditions of diversity and equality are elusive. One reason for this elusiveness, Garber (1995, 29) suggests, is a suspicion of common understandings that the notion of community implies. If difference becomes the normative window on local political life, then Garber is right in asserting that communities of choice and common understandings are almost always juxtaposed against communities of place. With this point, we return to the problem of conceptions that pose gesellschaft against gemeinschaft because gemeinschaft is always historicized to an earlier period and, as such, it becomes little more than an element of nostalgia that people try to regain. As I noted earlier, community established in this form historicizes and marginalizes community politics, along with local female activism, to "public housekeeping."

Young argues that, given the prevalence of this mythic conceptualization of community, justice is hindered at the local scale but it may not necessarily be hindered at the scale of cities themselves. Her alternative to the idea of community is "an ideal of city life as a vision of social relations affirming group difference" (1990b, 227). Young's vision remains somewhat utopian because cities tend to cultivate the worst forms of both communitarianism and liberal capitalism, but her point that we need to start with what we have is well taken. Affirming the liberating aspects of cities is a worthy goal on the road to realizing a justice that recognizes difference. The question that remains is, How can difference be recognized through critical regionalism?

One of the most poignant examples of this kind of critical regionalism arises from the single-mindedness of Alice, the single mother whose child was born with cerebral palsy. Although I am uncomfortable making any kind of generalization about women with "special needs," there seems to be a connection between single mothers like Alice and Carol in their ability to form community creatively out of institutional space. By maneuvering through the complexities of the local political economy and social culture, these women use their marginalized position to empower themselves and their children. As Alice will articulate (much better than I can) in a moment, her child's special needs enabled her to use their peripheral position at the edge of "normal" family life to affect change.

Alice did not consider any kind of support network prior to the birth of Mary-Jane. She had problems with the child's father during pregnancy. In our first interview, it was clear that Alice resented the father's not being around, but at no time did she talk about the possibility of asking for any kind of support from him or anyone else. "I am on an emotional roller coaster because he is acting like a horse's ass." Alice wanted to be in control. "Now, with the unplanned pregnancy, I'm going to have to stay in charge. If I had a choice, I wouldn't be having a child and also working in childcare."

Alice talks about being lonely prior to Mary-Jane's birth. She was new to San Diego, working eight hours every day and then baby-sitting at night to make ends meet. If Alice had any sense of place at that time, it was focused on the whole of San Diego rather than her local area:

> I'm still really impressed with San Diego freeways [laughing]. I can get to someplace within fifteen or twenty minutes. And when you work all over the place like I do, that is important. I barely drive on surface streets. It seems that wherever I'm going is not far from the freeway itself, it's just a short distance, in contrast with [my last residence in] L.A., where you have to drive for five or six miles [after the freeway], and then you're there. And then you've got to deal with parking problems. Everything seems really close to the freeways, and that really helps. It makes driving a whole lot less stressful; you don't have to go through some of the city.

Alice lives in Normal Heights, an uptown neighborhood developed in the 1930s on the northern edge of the old trolley system. Although the area still comprises a large proportion of single-family homes, a signifi-

cant number of apartment complexes now fill what was once a relatively low-density landscape. Alice lives in a small, two-bedroom apartment in a complex of six units.

When Mary-Jane was born with cerebral palsy, Alice was fright-
ened and lonely, but she realized that she and her daughter had special
needs. She slowly established around herself a tight network of support:
asking no help from family or neighbors, she first sought help from professional sources. As a result, her support network is not local but is
contrived from various regional institutions; she deliberately pieced together these sources of support to form what she calls a "lifeline" which
she can pull upon when feeling lonely and overwhelmed:

> Because of Mary-Jane's condition, I can access a lot of services
> which she qualifies for; one is Respite Care: that is an agency
> which provides care for special needs; you have to have a dis–
> ability or be elderly. They send people to your home. If I want
> to go on a job interview, shopping, a movie, or if I just want to
> go into the other room and watch TV, and I just don't want to
> take care of the baby, a social worker will see that I'm alone and
> say, "Yeah, she needs help." [Interviewer: Is this someone that
> would come every day?] No, they are not to come for, like, day
> care [Interviewer: Just to give you a "respite"?] Yeah, so today
> I had some shopping I had to do, and my roommate wouldn't
> be here! [Interviewer: How much notice do you have to
> give them?] About two days usually, they'll come right away
> in an emergency. If I know the address of the social worker,
> sometimes I can call them directly; but usually it is two to
> three days.

All networking is spatially and temporally contrived, but networks
limited to institutions often have restricted geographic access and hours
of availability. Alice's tenacity enabled her to break through the spatial and temporal boundaries of several large-scale institutions. She uses
words like "trustworthy," "supportive," and "flexible" to characterize her
imagined community:

> I heard about all these support agencies through the California
> Regional Center. They deal with all these kinds of handicaps.
> That has helped a lot. I have a teacher come to the house. She

gets [physical] therapy twice a day, either to my house or to where [the therapist] is. She does infant stimulation. There are a whole lot of things that impact my situation. Mary-Jane also goes to therapy twice a week. The key was getting access to all these agencies and getting into the system, which I was able to do at an early stage. Because if it hadn't been for them—all the different resources that I have access to, a lot of social workers, a lot of people that were able to be a network of support for me and Mary-Jane [voice trails of contemplatively]. There were people early on that helped me take care of that and all the mountains of paperwork I had to take care of. The [social security check for disabilities] helped financially, [but most important was] having people to talk to you about it. The baby's father is not involved at all, and my family isn't really involved; so they kind of took up the slack.

For Alice establishing a community required that she ask for help and that support existed in a flexible enough form to accommodate her needs. It took Alice some time to accept her own situation. The support she has come to rely on was by no means immediate, nor was it easy to access:

I think I'm finding out that San Diego, if you've got some kind of illness, San Diego is a pretty good place to be because they've got a lot of expert medical facilities—UCSD [University of California at San Diego], Scripps, Mercy—and I hear a lot of good things about these places; and then there's the Children's Hospital [where Mary-Jane was for two weeks]. They're all kind of right here (but you have to know how to look). My philosophy was shotgun, to get everything with a big net, get everything that's out there for Mary-Jane and pull it in, and then I could choose. I was able to use the system. My pregnancy was kind of la-di-da. I mean I was nervous about things, but I never expected Mary-Jane to be as ill as she was. I was shaken up and *alone.* You know, no family; the father was in another state; I was without a car, without a job. *What am I going to do?* I felt like I was caught off guard. I've got to run twice as fast just to stay in one place. [Interviewer: You feel much calmer now though?] Yes, I do! [Interviewer: Because you know these resources are available and they work?] Yes, I know I can pick

up the phone, and I can call [for support for just about any-
thing]. Even if I am just really scared, I can call a social worker
to talk to me or counsel with me, you know, give me some
things to help [in] dealing with the stress. A lot of helpful
people—I met more people in the first two months after Mary-
Jane was born than I have in my entire three years in San
Diego. And I don't know if I would have done that if Mary-
Jane hadn't been ill. [Interviewer: Sometimes it is hard to ask
for help.] Yes, it is, and now it's like no problem if I need [to
talk] or a drop-in. I do it!

An important aspect of connectivity and community for Alice is
also what she can give back:

[The] S.H.A.R.E. [program], we do that![4] That's great, we love
that. That was very wonderful to get involved in that: basically
it is very community-oriented; and it gets people into volun-
teering; and the time when I wasn't working, I became aware of
how important it is. There's a whole network of people who are
working towards improving family resources, and I think
S.H.A.R.E. is one of them; and I found connecting onto that
very empowering. It put me in contact with a lot of concerned
people who, you know, want to contribute to the community. I
think it is a wonderful program. . . . The American Red Cross
has an infant-nutrition program which is for pregnant and
breastfeeding women and infants: infants under the age of five,
I'm not sure. They give you vouchers that pay for all baby
formula. They pay for all of it, and when I was nursing, they
made sure I had food that was high in iron, calcium, and
Vitamin C; and when you stop nursing, they continue the
baby's nutrition with formula, and that is a great program.

The important point to stress about Alice, and also about Carol
and Doreen, is that part of their strength and ability to find support
comes from their image of uniqueness. It is not necessarily the system
that is providing for these women but rather their own sense of worth
and willingness to ask for help. Listen to how Alice tells it:

As awful as it was—and it was awful—in one way, if there is a
way to say that Mary-Jane's difficulties were a blessing, without
this I would be a struggling single parent with a deadbeat dad

[laughs]. But in another way, especially at my age [late thirties], regardless, you just prioritize. What has to get done gets done. I think we would have survived [without knowledge of the system and its resources; we would have] found things we needed, regardless. [Interviewer: Sounds like you've got a healthy network set up.] Yeah, and I'll never know what it would have been without my problem, what my choices would have been. [Interviewer: There aren't too many choices unfortunately.] No, there are not! I went from being afraid all the time, very afraid [laughs], very afraid. Even before the baby was born [and not knowing about the cerebral palsy], very, very fearful. And now I feel confident about my own mothering skill, but I [still] have also a lot of trepidation about Mary-Jane's problems. At least I know that I can take care of her.

The regional system is not what is central here but rather the use that these women make of it. In actuality, the system often turns out to be relatively impenetrable and unyielding. Here, for example, is Alice's account of her trouble with the welfare system:

I had the baby. I had to go through the welfare system. That was scary. That was very frustrating. That was getting up at five o'clock in the morning, taking a bus at five with the baby because it opened at 7 A.M., and you had to be in line by 6 A.M. They only take eighty people, eighty new cases or something like that. You had to be in line by [thoughtful pause]—I can't remember, but all I know is that it was *dark* . . . because there was a big line already; and then you were there until two—very eye-opening. There's no way the system can work unless you lie, unless you hide and lie; it's not designed to, and I guess they know that, so they don't give you information until you've done something wrong, until you've made a mistake; and then you're penalized for it, and you have to go through the whole thing again. It is very frustrating, and in the end I realized that I would rather struggle and go back to work than get caught in the system because it's like a black hole of despair. It was awful [laughs]. It is not designed for people who are working their way through. You've got to be completely, you know, flat broke, destitute. Basically, one step, one foot out of the door, or on the street. They kind of misrepresent that; they kind of make you think, "We'll help you; you will be okay. We can help you stay home with your baby." They don't help you do that!

Women like Alice can make the system work for their families even when faced with the bureaucratic complexity of gesellschaft. Their stories reveal women as active participants in reshaping their families' lives rather than dupes and victims of an unfair society. Although the contexts of these women's lives seem good, I do not want to soft-pedal the temporal and spatial constraints that they face. In addition, I think it is important to note that these women actively participate in the continuance of their communities. For example, Alice helps with S.H.A.R.E. and works as a day-care helper, and Carol manages an elementary school and volunteers with the Heartland Mothers of Twins Groups.

Scale and the Unoppressive City

Young (1990b) joins with feminist Elizabeth Wilson (1991) to highlight the important positive dimensions of city life for women. As we saw earlier, the works of fin-de-siècle theorists and some contemporary writers, such as Robins (1996), suggest that the contemporary city is an oppressive and enervating social structure. The ideal of Young and Wilson is "the unoppressive city," and they believe that such a city exists under veils of patriarchy and capitalism. "Even for many of those who decry the alienation, massification, and bureaucratization of capitalist patriarchal society, city life exerts a powerful attraction. . . . For many people deemed deviant in the closeness of the face-to-face community in which they lived, whether "independent" women or socialists or gay men and lesbians, the city has often offered a welcome anonymity and some measure of freedom" (Wilson 1991, 317). What the fin-de-siècle theorists saw as deviant behavior, then, is celebrated as difference and diversity in postmodern and feminist parlance. The scale of larger cities means more diversity, but it also means that diversity has more of a chance of being honored instead of being merged into communal norms or forced underground.

Thus, Young believes with Jacobs and Gans that large cities liberate people from conformist pressures, but she is skeptical of any kind of political value for difference at the scale of community or local governance. Rather, difference is constituted, experienced, and politicized through the infinitely unique spatial and temporal distinctions in cities (Young, 1990b, 318):

> The temporal and spatial differentiations that mark the physical environment of the city produce an experience of aesthetic *inexhaustibility*: buildings, squares, the twist and turns of streets and alleys offer an inexhaustible store of individual spaces and things, each with unique aesthetic characteristics. The juxtaposition of incongruous styles and functions that usually emerge after a long time in city places contributes to this pleasure in detail and surprise. This is an experience of difference in the sense of always being inserted. The modern city is without walls; it is not planned and coherent. Dwelling in the city means always having a sense of beyond, and there is much human life beyond my experience going on in or near these spaces, and I can never grasp the city as a whole.

The positioning of Young's arguments against a contrived notion of scale that is linear and oppositional is important. In the words of Smith (1992, 76), "It is [the naturalization of] scale that delimits the prison walls of social geography." For Young, the uncontrived scale of city life does not build walls, but, rather, it celebrates difference and diversity because it is about the being together of strangers. This togetherness is not necessarily intimate or face-to-face: it may be contrived by the media or the voyeurism of *flânerie*. Although people may remain strangers, they acknowledge "their contiguity in living and the contributions each makes to others" (Young 1990b, 318). They experience each other through difference rather than sameness: intervention, contradiction, and contestation are all possible.

The geographic form of Young's social justice challenges the liberal democratic model of small, decentralized, autonomous communities for urban decision making. Instead, she argues for a form of social justice with equality among groups that affirm one another in larger regional units with mechanisms for incorporating and representing smaller units and towns. This form of representation is predicated on empowerment without local autonomy. For Young, local autonomy is problematic because it implies sovereignty. Thus, with decentralized autonomy, having *sub*urban privatopias and small communities exercise control would mean that citizens in each municipality could decide on their own form of government, their own rules and laws, how their land and economic resources could be used, and so forth. Such local autonomy can only foster large-scale inequality among communities and thereby the oppression of individuals who do not live in more privileged or more powerful communities. A community that elects to have no bus stops within its boundaries, for example, denies access to those who need public transportation. At the between-group level, this form of social justice will also no doubt result in exploitation of one municipality by another.

The potential of Young's project is that it points to problematizing spatial scale as a social practice as well as a political act. It conceives of social and spatial processes as a multiplicity of actions and hierarchical structures that cohere and contradict, some of them exploitive and others liberating. Although Young's work neglects any critical appreciation of the social production of scale, nonetheless the solution that emerges potentially destabilizes hierarchical oppositions. Her regional solution may appear utopian, but it also provokes because it trivializes the "naturalness" of scale. As Marston (1995) points out, the connections between scales are not given naturally but are made. Young (1990b, 250) proposes to contrive spatial justice from a regional sovereign authority that denies local autonomy to powerful, small-scale communities and jurisdictions. "Where there are diverse and unequal neighborhoods, towns and cities, whose residents move in and out of one another's locales and interact in complex webs of exchange, only a sovereign authority whose jurisdiction includes them all can mediate their relations justly."

Young is producing a scale as a deliberate political act. Implicit within her analysis is the understanding that there is nothing ontologically given about the traditional divisions between neighborhoods, towns, cities, and regions. The geographical structure of social interactions establishes geographical scale, and, as such, her regional solution usurps the ability of wealthy, private communities to jump scale.

Smith (1992, 73) points out that a critical theory of scale suggests that we need to avoid both a relativism that treats spatial difference as a mosaic of unique local areas, and a reified and uncritical division of scales that repeats a fetishism of space. Clearly, Young's model avoids the first indictment, but it not entirely clear if it successfully negotiates the second. What remains unaccounted for in Young's regional solution, and why it appears utopian as a consequence, is a critical appreciation of the meanings of scales and of the contradictions and tensions among scales. A neglect of these issues raises the potential for a static formulation of scale. Smith (1992, 74) proposes three reasons why we need to be critical of solutions that suggest a static view of scale. First, we need to be able to interpolate the "translation rules" that allow us to understand not only the construction of scale itself but also the ways in which meaning translates between scales. Second, in that scale is used to give meaning to social interactions, we need to understand its metaphorical power. A critical theory of the production of scale would differentiate and integrate the meaning of scale as a metaphor for social relations and as a grounded material practice. Young's solution falls short because it considers scale only as a material practice. Third, a critical theory of scale must speak to the construction of difference. Although boundaries are continually forged and reforged in social practice, powerful individuals and groups continue to metaphorically appropriate space to establish the centrality of their own positions. Young's regional solution would make these actions increasingly difficult to contrive, but the processes of scale construction still remain hidden. Smith (1992, 78) would agree with Young's intent, but he would argue further that the goal of a politics of spatial justice is not only "to overcome social domination exercised through the exploitative and oppressive construction of scale" but also "to reconstruct scale and the rules by which social activity constructs scale."

Scaling Communities and Everyday Differences

Although I emphasize the "scale of community," different scales are important at different times, and the relative importance of different scales during any period is an empirical question (Marston 1995). The latter half of the twentieth century is a time, at least in the United States and Europe, when the autonomous private community has become an increasingly reactive political tool. Reactive political administrations often use the rhetoric of decentralized and autonomous communities to construct and legitimize various fiscally driven policies. The assumed fiscal and legal autonomy provided by "community-development block grants" and "neighborhood-watch" programs, for example, speaks to our fears while at the same time it whispers our fantasies. To paraphrase Lacey (1994), we like to think that we live in real, identity-fostering, caring communities, but part of the postmodern experience is precisely the fear that we live in a modernity that constrains communication and consensus. In this context, the rhetoric of community assumes particular power in the hands of government and other purveyors of influential social discourses. In Young's model, the contradictions within and between communities are embraced, and the moving foundations of identity are accepted as axiomatic. By recognizing the contradictions and differences within ourselves, we are able to enter into spheres of communication, and we are able to accept dissent. Young begins to construct a scale through which ideas of difference acquire or lose their distinctiveness. Hers is a space where groups, classes, and races can constitute themselves and recognize one another; it is a space within which Lefebvre's trial cannot be constituted as an ideological battleground where patriarchy always gains the higher ground.

Without denying the importance of creating environments of mutual aid, companionship, and understanding within which children can be raised and nurtured safely, I argue that our images of community and family need to be reconstituted in ways that do not exacerbate by occlusion race, class, gender, and sexual diversity. Acceptance of the difference within ourselves and an attempt to distance ourselves from mythic images of a "normal" family embedded within a particular community form perhaps constitute one way of accepting difference and diversity. To reiterate my arguments in chapter 5 for relating play and justice: displacement of the logic and reason that constructed a form of justice based on difference and hierarchy is required if we are to join in a dialogue that allows both consensus and dissent. Difference should not be used as a justification for building hierarchies, whether they involve adults and children or families and communities. A more liberatory form of justice affirms rather than suppresses social-group differences, but, in so doing, it must also break the walls built by an uncritical acceptance of hierarchical scale. Problematizing scale in the manner proposed in this chapter opens new ways of thinking about the embeddedness of families and communities.

The thesis of this book is that our understanding of the spatial relations within families, and between families and communities, is constrained by myths that do not account for the diversity of day-to-day, lived experience. These myths are circumscribed and constrained by an imagined family that closely resembles the nuclear-family form and an imagined community that is only slightly removed from the rural idyll of a century ago. These images are contrived in part by a naturalized view of hierarchical scale relations. I have tried to show how these images are inculcated in academic practice and how some feminist and poststructuralist writing is trying to unpack their patriarchal baggage. Part of this dislodging reveals that, unlike the myths that seem to guide us, the day-to-day existence of families and communities is fragile. I try to privilege the instability of family life in conversations with parents and caregivers as they continually construct new lives for themselves and their families in unnatural spaces. My concern is ultimately how people define their places in households and communities as their individual expectations, commitments, and concerns change.

Conclusion

Negotiating Complex Family and Community Spaces

This book is about the construction of two social imaginaries, the family and the community, and how these myths play out in people's day-to-day lives. I suggest that the inveteracy of these myths is in part a product of their ability to endure continuous trials by space and scale. Barthes (1972, 155) describes the power of social imaginaries such as family and community:

> For the very end of myths is to immobilize the world: they must suggest and mimic a universal order which has fixated once and for all the hierarchy of possessions. Thus, everyday and everywhere, man [sic] is stopped by myths, referred by them to this motionless prototype which lives in his place, stifles him in the manner of a huge internal parasite and assigns to his activity the narrow limits within which he is allowed to suffer without upsetting the world.

Spaces for families are an unnatural product of the stultifying gender norms of history and geography; they are lodged between, and help to construct, the scale of the body and the scale of the community. This seemingly normal set of associations establishes boundaries that, as Young (1990a, 315) points out, constrain and limit change. "If institutional change is possible at all, it must begin from intervening in the contradictions and tensions of existing society. No telos of the final society exists, moreover; society understood as a moving and contradictory process implies that change for the better is always possible and always necessary."

Some academics are attempting to pull together spatial and temporal aspects of patriarchy but, as yet, no social theory satisfactorily addresses why and how and when and where societies regulate this and other forms of social reproduction. This lack of understanding can be attributed in part to the complex variations among families, but, this aside, all societies regulate the coexistence of people and the rules of reproduction. I argue in this book that one reason for the absence of adequate theorizing about reproduction is a lack of consideration of spatial and hierarchical rules and regulations. In pursuing this deficiency, I make no claims to discovering any fundamental truths about how spaces and hierarchies circumscribe changes in households. The multiple voices and dialogues that I reproduce in this book do not point to any new universal revelations; some of the insights from our conversations with young San Diego families may appear intuitively obvious, while others are "nuggets" that affirm parts of my theoretical discussions. Through these conversations, a diverse and contradictory array of gender, cultural, and ideological resources are woven with people's perceptions and actions in a concrete, day-to-day, urban environment to graphically demonstrate the protean nature of family and community life.

Given my reticence to set down guiding principles for any regularities we may have observed, I believe that we need to question what is intuitively obvious because the obvious is usually a norm and intuition is often a social construction. Popenoe's (1993) suggestion that the family environment for child rearing and affection is failing because parents are more interested in individual rather than familial pursuits is not reinforced by any part of this book. Many of the parents we interviewed certainly express feelings of overwhelming spatial and temporal entrapment and stress, but the vast majority are passionate about being the best parents for their children. The problem is that most are working beyond their limits to provide their time and their emotions in spaces and communities that are often restrictive and disempowering. Rather than exhibiting a lack of interest in the process of family making as suggested by Popenoe and other communitarians and traditional family-values advocates, the men and women we spoke with struggle hard not only to make decent families but to do so in spite of great political, economic, social, and geographic barriers to the creation of a child- and family-focused life. It seems that the only support offered by Western society is a series of myths about how the family and the community should look. I argue that although society may be conceived as undergoing complex and contradictory processes of change, patriarchal notions of family and community are nonetheless reified by appeal to an ethereal, hegemonic legitimacy. In large part, their mythic structure constitutes, and is constituted by, the spaces of everyday life. The spatial performativity of the family and of community myths is evident in suburbia, but it perhaps finds a new, devoted proselyte in contemporary neotraditional communities. I argue that these are unnatural spaces for many people because they deny the differences between and within families. Assuming a mythic family norm, these spaces are unable to accommodate the fragility and uncertainty that are part of family life.

A crisis exists because residential space performs a choreographed dance that presupposes gender images and behaviors that are fantasies in the sense that most people find them unworkable. I have tried to show that family and community myths not only are unattainable but also are enervating in the face of change at the day-to-day level. Notions of the essential components of family life miss the complexities of daily living and often follow racial stereotypes. For example, Hispanic and African American families are often characterized as extended, but sometimes implicit within this characterization is the suggestion that the extended family is merely a less developed nuclear (that is, white) family. In fact, however, the examples that I highlight throughout the book suggest that most families are, to some degree, extended. This conclusion may seem facile, but it is complicated by the fact that these families are wrapped in complex social networks that may involve kith, kin, and institutional structures, and they are part of complex geographies that embrace gender negotiation, local connectivity, and critical regionalism.

The book illustrates that the boundaries imposed by our understanding of what constitutes space and scale create a tension between what is thought to be normative and what is experienced. The raising of children provides a consummate theme for my discussion of the experience of families, and I appropriate the term "community" to provide a context for my discussion of those families' geographies. The term "family" is powerfully etched in a politics of identity and space that contrives problematic and outmoded notions of motherhood, fatherhood, and childhood. It is my conviction that the arbitrary values of patriarchy contain and contextualize community in the same way that they direct our feelings about family. Like the family, the private, autonomous community has become an increasingly reactive political tool. Also, like the family, the reality differs from the image: we like to think we live in identity-fostering, caring communities, but part of our day-to-day, lived experience is precisely the fear that we are increasingly adrift in a complex and uncaring world. Nonetheless, a social imaginary is created around family and community insofar as there is social consensus about how they are constituted and insofar as there are generally warm and fuzzy feelings around our ownership of them. The problem with the spatial interaction and the scalar interpenetration of these concepts is that they rarely allow us to contest and play with the ways in which they are imagined.

The men, women, and children in our study may be creatively remaking American family life at a time when the very notions of family and community are under scrutiny and traditional gender relations are being criticized. In our conversations we found the notions of family and community being reworked in many new and interesting ways. We observed some families barricading themselves behind right-wing religious traditions and others searching for new ways of living in neotraditional, gated communities. In other cases, single mothers created community for themselves out of institutional contacts and critical regionalisms. We found that some men and women jumped scale in conscious and unconscious attempts to circumvent oppressive spaces and constraining ideologies. Certain working women struggled to involve reluctant husbands in domestic activities, and some enthusiastic fathers struggled with new issues of identity that surrounded their commitment to childcare.

The importance of undermining the naturalness of family, parenthood, home, and community is constituted in what Joe Kincheloe and Peter McLaren (1994, 154) call "critical world-making." As a process guided by the shadowed outline of a dream of a world less conditioned by misery, suffering, and the politics of deceit, critical world-making suggests a way forward. I see this project as a way to consider how fatherhood, motherhood, and childhood are being defined in allegorical space as well as hegemonic space. Today, academics and practitioners are being increasingly flexible in their constructions of people's daily life. Terms like "family" and "community" are still, to a large degree, monolithic, and they are certainly embedded in a naturalized conception of scale; but the practical implication of this embeddedness is perhaps less threatening than it was even in the mid-1980s. For the most part, we are no longer constrained by an arcane need to develop generalizable models of family and community life. It is generally agreed that the relations between family and community are complex and changing, and some academics are content not to try to mold this complexity and change it into something that constrains real-life experiences. The gender performances of mothers, fathers, and children in the spaces of day-to-day living are constituted by spatial power relations that circumscribe a choreographed constellation of imaginaries. It is important that we try to understand how these performances come together in a critical form of world-making that is not constrained by myths.

Rethinking the world anew is a particularly thorny problem for me as a geographer because the "naturalness" of space and scale are only now being questioned. The "geographic turn" in theorizing about social relations enables a discussion of scale and spatial performativity alongside feminist and poststructural critiques. Joining Lefebvre's project on how space is produced with the work of Butler on the performativity of that space enables me to highlight theoretically how the monolithic concepts of family and community endure. Reified and fixed in seemingly immutable configurations of knowledge, these myths legitimate a choreography that presupposes a "familiar" spatial grammar for motherhood, fatherhood, and childhood. Not only do contemporary family configurations not fit well into this dance to begin with, but they are also continuously changing and thus attempting to dance to different tunes. Moreover, most "family-oriented" communities are ill-equipped to accommodate these kinds of changes. It may be argued that the rigidity of community and institutional structures regulates family formation and makes it difficult to contest norms, and it seems to me that we need to question how the resistance of communities to change may be attributable, at least in part, to the apparent "naturalness" of the family. Certainly, if we describe families from within a rhetoric of kinship and home, then they may denote something to which we feel naturally tied. But Anderson (1991) argues that in everything "natural" there is always something unchosen. In this way, the family is not only connected with parentage but also with skin color, place of birth, and birth era; thus, the "naturalness" of the family exacerbates by occlusion differences in race, gender, and citizenship. In a grand essentializing manner, the naturalness of the family often appeared as a backdrop to our conversations with San Diego families—irrespective of race, ethnicity, or gender differences—and the myth's unworkability was just as often highlighted in the practical, day-to-day contexts of mothering and fathering.

Anderson goes on to argue that if something is natural, not only is it in part unchosen but it is also, as a consequence, "interestless." Thus, if the family is thought to be natural, it may be conceived of as a domain of disinterested love and solidarity. But if the family is "interestless," then, for exactly that reason, it can ask for sacrifices. These are the sacrifices made by parents, for example, to obtain an appropriate kind of house in the "right" kind of neighborhood. These are the sacrificed relationships when the commitments and responsibilities of family members change. These are the sacrifices asked of a single mother because she cannot command a family wage. In this way, arguments for the apparent naturalness of the family can lead to its use as an increasingly reactive political tool. Alternatively, if we position families within a tapestry of social and spatial constructions, then we need to account for new forms of community that justly reflect the diverse and continuously changing lives of men, women, and children.

Accounting for and interpreting change is a major proposition for current research on families. Contemporary notions of a monolithic family norm are disabling because they disregard concrete changes in the roles and relations of women and men and in the ways that child rearing is now practiced. By changing the contexts of their lives, mothers are placing new demands on the household, community, and society for employment, childcare, transportation, and social transactions. The roles of fathers in contemporary society are changing also, with gender responsibilities for domestic labor gradually being perceived to be interchangeable. The ways that aunts, uncles, grandparents, friends, and professionals participate in child rearing are also changing. These changes reflect a geographic and temporal agitation that makes it impossible to suspend and reify any notion of a universal, monolithic family as an existent reality. These changes point not only to the fragility of our construction of families but also to how the "idea" of a family continuously transforms itself as the commitments of its members change. Such transformations contribute to our understanding of the difference and diversity among families.

We are only just beginning to understand the potency of the hegemonic spaces of family and community as they destabilize around these kinds of changes. If we move for a moment away from the ideological bases for our hazy understanding of the anachronistic duo of productive and reproductive activities and move toward the grounded geography of day-to-day life, another form of spatial politics becomes apparent. This book demonstrates some of the ways that the monolithic concepts of family and community render a great injustice to the gendered and geographic complexities of raising children. Nonetheless, these concepts are pervasive in the sense that they are the concrete embodiment of somewhat nebulous values held by many people. Some of the stories in this book suggest that often those parents who tenaciously hold on to monolithic notions of family and community struggle most with their new family responsibilities. Reification of a family or community ideal is of little help in the face of change, and neither is the stigmatization of the increasing numbers who live in "alternative" family arrangements. In fact, "alternative" families at the "margins" often demonstrate new and healthy ways to redefine parenting, families, and communities.

Of concern is the continued standardization and naturalization of neighborhoods that are submerged in a family ideal that is no longer tenable. Geographers and planners should be directing public attention to legal, economic, planning, and social-policy reforms that could facilitate the coping mechanisms of young families. What can be done to reduce the prevalence of families' distress and to increase the adaptation of contemporary families through restructuring work schedules and benefit policies; redistributing work opportunities; enabling both women and men to earn a family wage; providing universal health insurance, prenatal care, perinatal care, and childcare; revitalizing public education or removing geographic impediments to appropriate education? Not all geographers or family researchers are competent to design political policies or planning trajectories or to suggest social services that might increase the stability and richness of family life. I know that I am not! But all who study families and who turn to empirical data on families to support their theoretical suspicions about contemporary family life need to highlight ideological motivations. We need to make certain that the research we do—the questions we ask and how we interpret the answers—adequately addresses the complexities of contemporary family life. We need to be sure that our agendas do not reify one strategy rather than another. Today, we craft a complex multiplicity of family and household arrangements, and we inhabit uneasy relations with our local neighborhoods, reconstituting ourselves frequently in response to changing personal, occupational, and community circumstances. Highlighting the spatial and hierarchical construction of everyday family life in the way that I have here does not lead to a new model for family life, nor does it reconstitute the role of community. Rather, it unpacks some weighty patriarchal baggage and focuses attention on the enervating qualities of spatial power relations.

Much of my own experience as a parent touches deeply on how this book comes together. The partial and painful peeling of my own images of family life were translated into my research in critical ways because it required me to reconstruct my perceptions of the world anew so that I could see more clearly the unnaturalness of parenthood. My continuing struggle is to open and pry apart the spatial politics of families and communities. This project is incomplete in the same sense that my relationships with Peg, Ross, and Catherine are incomplete.

Appendix A   Map of Study Area

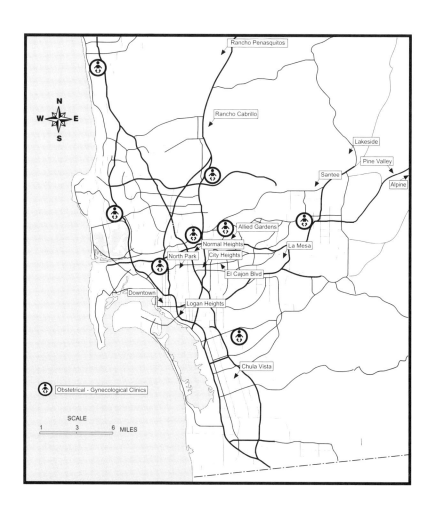

Appendix B

Selected Summary Statistics for Study Respondents

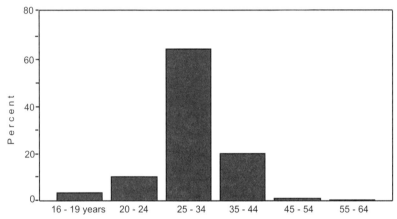

Figure B.1. Age (valid cases = 572; missing cases = 5)

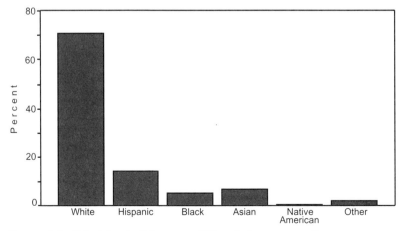

Figure B.2. Ethnicity (valid cases = 572; missing cases = 5)

Appendix B

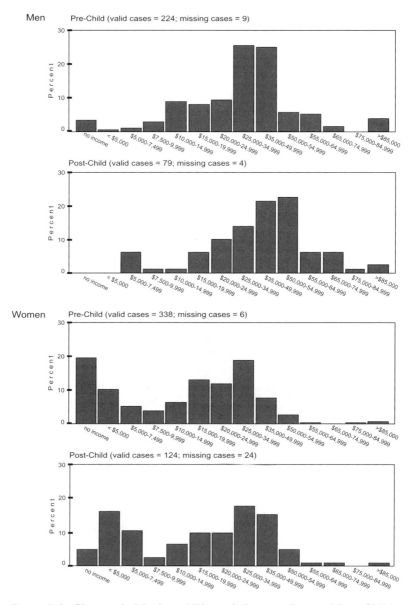

Figure B.3. Changes in Men's and Women's Income (Pre- and Post-Child)

		Pre-Child				Post-Child				
		Paid Employment Status, percent	Work Hours per Week	Hours Spent at Home per Day	Commute Time (minutes)	Paid Employment Status, percent	Work Hours per Week	Hours Spent at Home per Day	Commute Time (minutes)	Time Off to Have Child (months)
	Men	Full 90 Part 4 None 6	45.3	5.7	23	Full 91 Part 6 None 3	46.4	5.9	23	0.5
	Women	Full 51 Part 18 None 30	36.5	6.4	19	Full 43 Part 21 None 36	34.9	7.0	20	3.2
First Child	Men	Full 91 Part 2 None 7	45.2	5.3	21	Full 89 Part 8 None 3	47.0	5.4	26	0.6
	Women	Full 59 Part 14 None 27	36.5	6.1	20	Full 50 Part 20 None 28	37.2	6.8	23	3.6

Figure B.4. Changes in Employment (Pre- and Post-Child)

Appendix C

This appendix comprises brief profiles of the participants whose stories are outlined in some detail throughout the text.

- Although she is currently a single mother, Doreen was married to Alonso when first contacted. She lives in a single-family home in North Park, a mixed-race multi-ethnic community located in San Diego's Uptown (about four miles west of the Central Business District). She is in her early thirties and earns just over $25,000 per year. Alonso and Doreen separated when their son Scott was about six months old. Up until our second interview, Alonso lived in North Park but then he moved to take a job in Los Angeles. Doreen is white and Alonso is Hispanic (he was born in Spain).
- Russell and Trisha are a young white couple in their early thirties. When we first contacted them they lived in Santee, a peripheral suburban/semi-rural community located about fifteen miles west of San Diego. Just before Savannah was born they moved to a more peripheral location in Lakeside. Although both Santee and Lakeside are rapidly becoming commuter settlements they still have a strong "rural" and, in some places, "redneck" feel to them. Russell and Trisha live on Russell's income of around $90,000 per year. Between our second and third interviews the family broke up with Trisha and Savannah moving back to Trisha's parents in San Francisco. Our third interview was conducted at a time when Trisha returned to San Diego for family counseling. When we last talked, she and Russell were together again and had moved to a community 60 miles to the north of San Diego.
- Just before their daughter was born, Cindy and Paul moved to Rancho Cabrillo, a relatively exclusive neighborhood located twelve

miles to the north of downtown San Diego. Paul declined to participate our study but we know that the family is able to live comfortably
on his income. Cindy stays at home to raise their daughter, and is
president of the local Las Madres group. Both Cindy and Paul are
white.

- Tatiana is a young African American mother who lives with her
  mother and grandmother in an apartment complex in a run-down
  neighborhood in Chula Vista, seven miles south of San Diego's
  downtown. Tatiana is enrolled in a professional development degree
  at a local community college. She hopes to marry her boyfriend once
  she has completed her course-work.

- Prior to the birth of Hannah, Allen worked in construction and as a
  teacher at a special education school. When we first met him, his
  combined income with his wife Janet was about $70,000. This
  dropped to $45,000 when Allen decided to give up his jobs to stay at
  home with Hannah. Between the second and third interviews they
  had another child and Allen remained at home. They live in Santee
  in a fairly high density working-class neighborhood that borders
  Mission Trails Park (a regional park that extends over several small
  mountains). The neighborhood was described as "ram-shackle" in the
  field-log of one of the interviewers. Their house is a modest
  three-bedroom detached single-family home.

- Barbara and Peter live in Allied Gardens, a suburban community
  developed in the 1950s: Both their families live close-by. Barbara
  works for a marketing company and spends every other week traveling. After the birth of the twins, she was able to work out of her
  home on the weeks that she was not traveling. Peter works for a Navy
  contractor on a fairly regular nine to five job. They have a combined
  income of just under $70,000.

- Alice is an African American single mother in her late 30s. The
  father of her child, Mary-Jane, left before the before the birth.
  Mary-Jane was born with cerebral-palsy, a neuro-motor disorder that
  usually occurs because of delivery problems in about 1 in a 1000
  births. Alice lives in Normal Heights, a mixed income neighborhood
  located six miles to the north of downtown San Diego. Alice lives in
  a small two-bedroom apartment in a six-unit complex close to Adams
  Avenue, the main thoroughfare of the area. At our first interview,
  Alice was a employed as a pre-school worker but she gave this up
  when Mary-Jane was born. She earned less than $20,000 a year. At

our last interview Alice was actively looking for another pre-school job.

- Sandra Sanchez is Chicana, living with her "extended" family that, during our first set of interviews, comprised her father Benito, her mother Maria and uncle Roberto. The family originally migrated to New York from Puerto Rico in 1973. They own a house in Logan Heights, a primarily Chicano community located southeast of downtown San Diego. At our first interview, Roberto and Maria's income provided the household of four people with just under $35,000. When he left for Los Angeles, Roberto still sent some money home, but his position as a financial provider was only partially taken up by a brother and Sandra's boyfriend who became part of the household. At our last set of interviews, the combined income of household (which now had six members including Sandra's new baby) was just over $25,000.
- Carol lives in Lakeside with her twin boys. At the time of our last interview, her husband Will was filing for divorce and was permanently stationed with his ship in Long Beach, California. She earns over $50,000 from her position as a manager of a private elementary school facility where she works between forty and fifty hours per week. She lives in an apartment complex in Lakeside within three miles of both her grandparents who help look after the twins when Carol is working.

Notes

Introduction: Encountering Family Fantasies

1. A spatial comparison of the interview sample with San Diego census data suggests that the majority of those interviewed have zip codes associated with a slightly lower than average income, a slightly higher than average incidence of single-parent households, and a higher percentage of the population living in the same house over the previous five years. Beyond these statistics, the interview sample is not significantly different from the general population of San Diego (Herman 1995). Summary statistics for the whole interview sample are presented in Appendix B.

2. The attrition rate between mail surveys averaged over 30 percent, but, for the duration of the study, we maintained over half of the respondents who agreed to participate in the in-depth interviews. The bulk of the empirical material in this book comes from a selection of the repeated in-depth interviews of adult household members over the three-year period.

Chapter 1   Re-placing the Family: A Space for Difference

1. Many interview transcripts were analyzed through theoretically informed critical interpretations. Throughout the book, I make use of quotations, some of which are quite lengthy and separated from the text, to allow the participants to describe the changing contexts of their families and communities. The quotations are presented with as little editing as possible. Discontinuous material is indicated by ellipses (. . .). Square brackets [ ] are used for interjections, comments, and questions by the interviewer and for changes in, and additions to, participants' words where necessary for continuity or clarity. Square brackets are also used to describe body language or emotional outbursts, and words that are emphasized by respondents are *italicized*. Participants are identified by aliases or are left unidentified to preserve their anonymity. The stories of nine families are highlighted to varying degrees throughout the book, although I select excerpts from many of the other interviews as they pertain to my theoretical arguments. If unclear from the text, some excerpts are followed by a characterization of the speaker (mother, father, an uncle). Appendix C offers brief profiles of the nine families whose stories I elaborate.

2. The locations of neighborhoods mentioned in the book are indicated on the map in Appendix A.

3. Evelyn Glenn (1987) points out that anthropological literature is replete with examples of configurations that differ from the conjugal family but nonetheless

fulfill basic domestic and child-rearing needs. She notes that even Murdock conceded that the actual configuration of the conjugal family varied greatly, being most often embedded in a larger domestic unit or community.

4. A particularly problematic concern is how the economic base can be defined without having recourse to categories that are themselves superstructural. Most contemporary Marxists emphasize the impossibility of splitting social, cultural, and economic analyses at any theoretical level.

5. Glenn (1987, 358) points out that because feminism is rooted in personal experience, oppression is not treated as monolithic. Thus, for feminists the family is oppressive but individual families may not be!

Chapter 2   *Family Fantasies: Mythic Histories and Geographies*

1. I use the term "myth" as Roland Barthes (1972, 143–58) did when he suggested a process whereby a concept becomes naturalized and, in effect, depoliticized:

   In passing from history to nature, myth acts economically: it abolishes the complexity of human acts, it gives the simplicity of essences, it does away with all dialectics, with any going back beyond what is immediately visible, it organizes a world which is without contradictions because it is without depth, a world wide open and wallowing in the evident, it establishes a blissful clarity: things appear to mean something by themselves. . . . Every myth can have its own history and geography; each is in fact a sign of the other: a myth ripens because it spreads. . . . It is perfectly possible to draw what linguists call the isoglosses of a myth, the lines which limit the social region where it is spoken. As this region is shifting it would be better to speak of waves of implantation of the myth.

2. In Europe's American colonies, for example, the average household was larger, having six to seven persons. This difference is attributed to greater wealth that was more equitably distributed. More land was available in rural areas and healthier living conditions increased birthrates and decreased deathrates. In New York and other large urban areas, family size more closely paralleled that in Europe.

3. Lutz Berkner (1972, 1975) was the first to point out that Laslett's work was based on census materials that neglect change within families. Studying premodern families over time reveals in a significant number of cases, at least in the regions studied by Berkner, that grandparents lived with the conjugal unit. These generational or vertical extensions—known as stem families—usually had a sizable amount of property and wealth. Although not numerically overwhelming, this family form was common in southern France and parts of Germany and Austria. The notion of a family life cycle is important because the stem family would break up, subdivide, and rejoin over time. Berkner also found numerous examples of brothers banding together with their families at times of economic austerity.

4. Although analyses of early census data in England suggest a nuclear family norm, Miranda Chaytor (1980) shows that considerable diversity went unreported officially. Significantly, the size and composition of premodern families were likely to be affected by sudden death. Chaytor shows, for example, that some of the households in the parish of Ryton, County Durham, consisted of "hybrid" families of parents, stepchildren, and foster children as well as the offspring of the current marriage (see also Wrightson 1982).

5. During the Scottish Highlands clearances in the late eighteenth century, for

example, vast amounts of land were set aside for sheep raising and thousands of crofters were expelled from their tenant holdings. These clearances resulted in the depopulation of the Highlands, an area that remains relatively devoid of people to this day.

6. The transformations are well known: rapidly growing, small, independent cottage industries were increasingly consolidated into large-scale factories; farmers were dispossessed of their means of subsistence by the consolidation of their small holdings into large enterprises that were capital-intensive rather than labor-intensive; the wholesale clearance of tenant farmers from large tracks of land contributed to the growth of a large pool of labor that migrated to work in the new urban factories. According to Mackenzie and Rose (1983, 161), these transformations coincided with the creation of surplus value and the setting in motion of a self-expanding capitalism that creates (and re-creates) the production of surplus value in addition to value, the accumulation of profit and its reinvestment for future profit, and the continual separation of workers from what they produce.

7. Amy Swerdlow and her colleagues (1989) note that this exclusion of women and children from public life did not originate with the factory acts in the middle of the nineteenth century but occurred first sometime before in the wealthy bourgeois families of Europe and the United States. Originally, the owners of capital were the merchants and middlemen who invested in commerce, mining, and textiles. With industrialization, they put some of the wealth they had made into the new factories and transportation systems. All these sites of investment were outside the home and were male-dominated. In time, these merchants became the new upper class, realizing the power and wealth of the old aristocracy. As their fathers, husbands, and brothers gained power and wealth in the public sphere, mothers, wives, and sisters gained control over the domestic functions of consumption and became guardians of the reproduction of middle- and upper-class work ethics, at least in the limited sense of values because education and health were increasingly controlled by the public sphere. In addition, an ethic of "conspicuous consumption" developed with the tendency of wealthy men to demonstrate that they could provide for their women by employing numerous servants, now considered hired help and commodities rather than family members.

8. Mark Poster (1978, xviii) points out that Marx's concept of the working family as a unified class leaves unaccounted for the notion that individual workers may have interests in dominating women and children. Family historians, Marxists and other social scientists miss this point if they view the family as a unitary phenomenon, unchanging in its life cycle, and undergoing some kind of linear transformation through time.

9. Farm households worked in conjunction with neighboring households, and most of everyday life was experienced communally. Everybody in the village knew the business of everybody else: rumors of premarital sexual relations, adultery, or blasphemy could lead to hearings before church courts that had the power to fine, reprimand, and humiliate. Also, although a meeting of villagers in premodern Europe was usually a meeting of households represented by the patriarch, it did not necessarily exclude women, who often headed households.

10. The paradox, of course, is that emotional separation and individuation through the late nineteenth century and into the twentieth century do not reflect day-to-day life in the same way that they buttress society's mythic ideas about what constitutes the family. Thus, as women entered the paid labor force and some

chose not to become mothers, they were blamed not only for maternal depri-
vation but also for the overprotection of children. A dominant ideology surfaced and joined with Freudian psychological theory to blame mothers for any
failings in children, even if they had no children.

Chapter 3   Gendered Parental Space: The Social Construction of Mothers and Fathers

1. "Essentialism" ascribes immutable qualities on the grounds of sexual difference,
   and "naturalism" maintains that such qualities are natural rather than socially
   constructed (Kobayashi 1994). More recently, Chodorow (1994) refutes the idea
   that less individuation is a natural and essential category of womanhood.
2. Eggebeen and Uhlenberg (1985) use decennial census data to calculate the
   amount of time men as a group spend in family environments with children.
   Later marriage, a decline in fertility, and increased rates of marital dissolution
   during this period contribute to the sharp decline in paternal involvement.
3. Laqueur has been at the forefront of this debate since the publication of his
   book *Making Sex* (1990). In this painstaking account of sex differences, he uses
   Foucault's (1980) "genealogical method" to investigate the biological sciences'
   historical constructions of men and women. Genealogies are "constituting" processes that refer to historical knowledges, struggles, reversals, resistances, and
   strategies of domination that are constantly changing (Rosenau 1992, 67).
   Laqueur points out that contemporary biological definitions of sexual difference have as much claim to truth as the now culturally flawed theories devised
   by Aristotle or Freud.

Chapter 4   Negotiating Gender Roles and Relations around the Birth of a Child

1. Our initial set of mailings procured a sample of 571 individual household members. Of these, 231 returned our questionnaires one year later. This high attrition rate was expected from previous longitudinal studies and an unsettled
   California economy that resulted in a significant increase in residential and
   workplace mobility in the mid-1990s. The analyses in this chapter were of initial and follow-up surveys from the 231 individuals who remained in the study.
2. The important point about the classification of domestic activities in the discussion that follows is not the specificity of the tasks but rather the changes in
   how adult household members feel about these tasks before and after the birth
   of the child. Following Michelson (1986, 1988) and Nicky Gregson and
   Michelle Lowe (1993), rather than opting for exhaustive coverage, we focused
   our survey questions on major domestic activities that require considerable regular (weekly or daily) labor and time and, therefore, are the most important activities for allocating individual households to a particular form of the domestic
   division of labor.
3. For most respondents, provision of household income implied paid employment.
4. No interpretation of the data for post-child unemployed men is possible because of the small sample size (only 3 percent of the men in the second survey
   had no paid employment).
5. All respondents who indicated on our questionnaire that they were first-time
   parents were invited to participate in our in-depth interviews. Of the 217 respondents expecting a first child who returned the initial survey, 127 agreed to

be interviewed and, in the follow-up, 81 of the 99 people returning a second survey agreed to be interviewed.

6. On average, household respondents expecting a first child were four years younger than those who already had children.

Chapter 5 *Play and Justice: Placing Children within the Patriarchal Bargain*

1. Ariès (1960, 412) also suggests that education was the great event that heralded a move toward the recognition of childhood.

2. Focusing on changes in the formation of families, Postman (1982) notes that, beginning in preindustrial Europe and continuing through modern Western society, marriage often heralded full membership in the adult realm, but this transition has become less identifiable since World War II. Similarly, Ivar Frønes (1994, 152) argues that in the 1950s in Western society, early marriages (made possible by housing subsidies and rising standards of living) and restricted access to sexual relations produced a situation in which (for men) entry into the workforce, marriage, and a full sexual life all occurred within a relatively short period, indicating a "natural phase" for the passage to full adulthood. Such contexts no longer converge to produce an identifiable distinction between childhood and adulthood, and, in addition, presently a long period is needed to meet the formal qualifications for entering the workforce. As a result, clear-cut cultural divisions between childhood, youth, and adulthood are evaporating.

3. Detailed arguments for the power of representation through film media and the effect on our personal geographies are made in *Place, Power, Situation and Spectacle* (Aitken and Zonn 1994). In particular, a chapter I cowrote with Leo Zonn focuses in part on the gendered construction of children through family-oriented movies (Zonn and Aitken 1994).

4. Some feminist writing is critical of the marginalization of children as the other because of its similarities to the marginalization of women and minorities (Thorne 1992, Alanen 1994, Aitken 1994, Jensen 1994). These feminists are beginning to point out the deeply "adultist" nature of research that does not consider children as persons in their own right or competent actors in ongoing social life. Alanen (1994, 29–31) notes that, for the most part, the social sciences accept a contemporary, Western (middle-class, male) construction of childhood as the essential childhood, unchanging in space and time. This enduring focus produces not only descriptions of childhood but also prescriptions for its universalization and globalization: "The institution of childhood organizes for its inhabitants (children) a particular location and along with it, a particular range of experiences of what contemporary childhood is. It also organizes for children a view of, and knowledge about, the social relations within which they live." Alanen goes on to argue that keeping children and their standpoints invisible in feminist research helps also to keep a number of women's problems invisible both in theory and practice, and so it is in the best interests of feminism to start a dialogue on the "child question" as well as the "woman question."

5. In my work with Herman (Aitken and Herman 1997), we suggest that—unlike structuralists such as Jean Piaget, Freud, or Jacques Lacan—Winnicott does not problematize the separation of the child and her external environment primarily in terms of self-discovery, objective distancing, naming, rationalizing, or compartmentalizing. Rather, Winnicott describes the place of play and child

development as transitional spaces, which, we argue, bear a close resemblance to the ideas that surround Lefebvre's trial by space. Winnicott sees the separation of the child from her environment as a fluid, recursive process involving intuition, experimentation, and play. Winnicott's principal concerns are how children (and adults) bridge the gap between egocentrism and recognition of an external world and how they negotiate and renegotiate the relations between self and other. Rather than seeking fundamental categories of relations, he illuminates infinite possibilities for personal development by attempting to describe, at least in part, the creative processes through which individuals establish perspectives that reconcile the inner reality of the self with the external reality of society.

6. Winnicott describes certain objects such as stuffed bunnies and security blankets as part of transitional space because they may constitute an object of experience that is neither self nor mother.

Chapter 6 *Setting the Nuclear Family Apart*

1. I use the term "suburbia" to refer to a subset of the *sub*urban processes that occur on the periphery of urban areas. Thus, *sub*urban processes may include infilling and gentrification in inner cities if those residential developments are exclusive, bounded, guarded, and private.

2. Writing about English suburbs, Wilson (1991) points out that the mid-nineteenth-century suburban home was a new version of the English castle, with elaborate ideals of comfort and consumerism. "The wife and mother in the prosperous suburbs of Bayswater or Edgbaston was hedged about with rules of etiquette and dress, forced to take charge of a large household of dependents and servants, and often, of course, had to maintain an appearance of gentility on an income insufficient for the task" (1991, 45).

3. If American society was suspicious of Howard's notions of community, his ideas on differences between males and females found favor because they were conflated with an ideology that separated culture from nature. Howard saw the town and country as magnets whose positive valences could be combined in a union of country and town that avoided negative characteristics. His conceptualization of country as nature and feminine and of town as culture and masculine exemplifies common ideals of turn-of-the-century Western society.

   Human society and the beauty of nature are meant to be enjoyed together.
   The two magnets must be made one. As man and woman by their varied
   gifts and faculties supplement each other, so should town and country. The
   town is the symbol of society . . . of wide relations between man and
   man, . . . of science, art, culture, religion. . . . The country is symbol of God's
   love and care of man. All that we are and all that we have come from it.
   Our bodies are formed of it; to it they return. We are fed by it, clothed by it,
   and by it we a warmed and sheltered. On its bosom we rest. Its beauty is the
   inspiration of art, of music, of poetry, . . . but its fullness of joy and wisdom
   has not revealed itself to man. Nor can it ever, so long as this unholy, unnatural separation of society and nature endures. Town and country *must be
   married*, and out of this joyous union will spring a new hope, a new life, a
   new civilization. (1902, 17–18, original emphasis)

   The conceptual slippage between man and culture and woman and nature also influenced the expansion of the western frontier expansion before the turn of the century (Norwood and Monk 1987; Aitken and Zonn 1993). It created a nature that needed to be conquered and controlled but that also offered Eden-like serenity and closeness to God. This pioneering ideology, which positioned men as conquerors of the vast American wilderness, was later reconceptualized in a new kind of homesteading at the periphery of urban culture and within the countryside (a tamed version of nature).

4. Uniform white occupancy was ensured by restrictive covenants such as the following, which the owner had to sign on receipt of the deed: "That no part of said tract shall, at any time, nor shall any interest therein be leased, sold, devised, conveyed to or inherited by, or otherwise acquired by or become the property of any person whose blood is not entirely of the Caucasian race" (Declaration of Restrictions, all lots contained in Rolando, Unit No. 4, Union Trust Company, May 1947, San Diego, California). This was one of only two restrictive covenants for houses bought in this particular neighborhood, which lies two miles south of San Diego State University. The second restriction pertained to the distilling of elicit alcohol.

5. The exclusionary nature of American residential neighborhoods is a predominant theme of urban literature in the latter part of the twentieth century. Even general planning requirements that are based on measurable criteria such as low population densities and large lots and house sizes in fact result in population homogeneity. Spatial metaphors such as deconcentration and decentralization are frequently turned into empirical measures that quantify the extent of suburbanization within metropolitan areas. Despite some growth in nonmetropolitan areas in the 1970s and 1980s and an increasingly positive image of inner-city residential areas for white, middle-class professionals, since World War II, peripheral suburbs have become the dominant settlement type in the United States. Today more than three-quarters of Americans live in metropolitan areas, and more than two-thirds of those live in peripheral suburbs.

6. El Cajon Boulevard was the main route east out of San Diego prior to the building of Interstate 8 in the 1950s (see appendix A). El Cajon is a four-lane, and in some places six-lane, road that establishes a significant barrier between neighborhoods to the north, which are primarily middle-income and white, and those to the south, which are increasingly minority and low-income. Commercial businesses along part of El Cajon Boulevard tend to be low-rent establishments such as VCR-repair shops, pawn shops, and used-car dealerships. Prostitution is also common along some parts of the boulevard.

7. La Mesa is an older peripheral community that is now wholly encompassed within metropolitan San Diego. Between it and Lakeside are the larger communities of El Cajon and Santee. La Mesa, El Cajon, and Lakeside were linked to the core of San Diego by a light-rail trolley system in 1993.

8. Alpine is a small town in the Cuyamaca Mountains east of San Diego; like Pine Valley, it is now a dormitory community for San Diego.

9. Numerous time-space and journey-to-work studies, beginning in the 1980s, document the travel constraints that women, particularly mothers, experience (Tivers 1985, 1988; Michelson 1986; Hanson and Johnson 1985). Some of these studies reveal that women tend to travel less frequently, have less access to private transportation, and, concomitantly, have less access to services and paid employment. The journey-to-work literature is filled with evidence that, for the population at large, women work closer to home. In 1983, the average work-trip distance for women in the United States was 8.3 miles compared with 11.2 miles for men, and, in Canada, the average time spent commuting to work was

27.5 minutes for men and 22 minutes for women (cited in Johnston-Anu-monwo, McLafferty, and Preston 1995). From the aggregate sample of 577 individuals in San Diego, male occupants of households spent twenty-three minutes commuting while their female counterparts spent nineteen minutes. Six months after the birth of a child, the commute times for men in these families did not change, but women on average spent a little more time commuting (see figure B.4). In the San Diego study, this relatively insignificant change may be attributable to women's job insecurity after the birth of the child: a decline in Southern California's economy from 1990 onward may have meant that some women had to commute further distances when they returned to work after giving birth. Less than 60 percent of the women we interviewed were guaranteed the same job when they returned to the workforce.

10. Although neotraditional developments may be considered a small part of CIDs in the United States, their popularity is growing. Most commonly associated with the "innovations" of architectural teams such as Duany and Plater-Zyberk in Miami, Florida, and Calthorpe in San Francisco, these kinds of developments—with their pedestrian orientation, high densities, and walled exteriors—are increasingly found elsewhere in North America as well as in Britain, Germany, the Dominican Republic, and Turkey (McCann 1995). They may well represent the pinnacle of privatopian aesthetics.

11. McKenzie (1994) argues quite convincingly that CIDs are not only the prevailing norm but the future of American housing development. They promote homogeneity because CID buyers as a whole reflect about the same level of diversity in age, race, and income as homeowners generally. CID purchasers most often are an adult male and female with children (or empty nesters), and they are whiter and wealthier than the general population.

Chapter 7   *Imagined Communities*

1. Communitarianism is sometimes narrowly defined as membership in a communistic community or as a belief in such a community. The term is used more broadly here as a belief in a society that is organized by communities and as a basis of self-identity (see chapter 8).

2. Smith and Katz (1993, 67) note that we have to "look back to the fin de siècle to find an equivalent period in which space was comparably 'on the agenda.'" Marshall Berman (1982, 36), in a stirring reflection on modernism, argues further that the fin-de-siècle theorists may have had a better grasp of the postmodern than we think:

> A century later, when the processes of modernization have cast a net that no-one, not even in the remotest corner of the world, can escape, we can learn a great deal from the first modernists, not so much about their age as about our own. We have lost our grip on the contradictions that they had to grasp with all their strength, at every moment in their everyday lives, in order to live at all. Paradoxically, these first modernists may turn out to understand us—the modernization and modernism that constitute our lives—better than we understand ourselves. If we can make their visions our own, and use their perspectives to look at our own environments with fresh eyes, we will see that there is more depth to our lives than we thought. We will feel our community with people all over the world who have been struggling with the same dilemmas as our own.

3. For Weber, specialization and capitalism were good. As part of this specializa-

tion, he rejects Tönnies implicit insistence on the "universality of man" but notes—with his infamous concept of the "iron cage"—that bureaucratization is a depersonalizing agent.

4. Durkheim introduced the concept of "anomie" to describe this rootlessness in contemporary social life. Anomie is a social condition characterized by instability, the breakdown of social norms, institutional disorganization, and a divorce between socially valid goals and the availability of means for achieving them.

5. Spurred in part by the work of Butler (1993), geographers have engaged in considerable discussion about how lesbians and gay men contrive and create queer spaces. A discussion of this important work is beyond the scope of what I want to do here, but it is worthwhile noting that queer space, if it appears at all, is masked in suburban areas by symbolism and behaviors that are hidden from the heterosexual gaze, whereas in inner cities certain neighborhoods gain their identity through overt queer symbolism (compare Valentine 1993; Bell et al. 1994).

6. In the context of my arguments, it seems appropriate to emphasize Webber's gender-biased language.

7. Roberts and Glasser were certainly aware of academic work on community when they wrote their autobiographical and semiautobiographical accounts of working-class neighborhoods. Glasser, for example, was educated in Cambridge and actively campaigned "against the destruction of traditional communities and their tried and tested values." Before he wrote *Growing Up in the Gorbals* (1986), his celebrated account of life in one of the worst slums in Europe, Glasser lived in a remote mountain village in Calabria, and from this gemeinschaft experience he wrote *The Net and the Quest* (1977), which was filmed by the British Broadcasting Corporation. Nostalgically resurrecting images of community "as it used to be" in small rural villages or working-class communities is what Michael Smith (1984, 124) refers to as "selective recall."

8. The Chicago School did not necessarily endorse working-class communities as inherently good. Harvey Zorbaugh's *Gold Coast and the Slum* (1929), for example, suggests that ghetto environments destroy social networks and create lonely social misfits.

9. Robert Everhart's (1983) exploration of working-class culture and resistance in an American school suggests that Willis's themes may cover a large part of growing up in capitalist system, and Susan Ruddick's (1996) exploration of homeless youth in Los Angeles provides further evidence of resistance and capitulation under the global restructuring of capitalism.

10. East San Diego (sometimes known as Southeast San Diego) comprises many racial and economically mixed neighborhoods that, like "South Central Los Angeles," are often lumped into one crime-ridden area by local media and folklore.

Chapter 8   *Difference and Justice: The Place and Scale of Community*

1. I borrow the term "critical regionalism" from Ruddick (1996), who uses it to describe how the image of Hollywood—with its focus on young stardom and rising from the gutter—intersects in critical ways with the political identities of homeless youth in the area. I use it here in a slightly different way, focusing on women's abilities to rise above local spatial and temporal constraints by jumping scale to garner support from regional institutions and networks.

2. As I pointed out in chapter 7, this totalizing impulse is precisely what Webber

(1967) used to argue for his notion of places without propinquity and for the possibility of a homogenous, middle-class society.

3. Wilson (1991) points out that these relationships are more apparent in the denser central cities, where single mothers and women increasingly find themselves. Arguments for local community empowerment through urbanization open up the constraints of the earlier theorizing of Tönnies and Durkheim, who postulated that the density and heterogeneity of urban residential areas would move political activism and the construction of the self to the workplace. The distrust of feminist theory for these works stems from their advocating rational social structures and the notion of people as passively responding to larger-scale social and economic change. Structuralists might argue that grass-roots political activism got its greatest impetus from economic and social restructuring around the demise of Fordian-Keynesian economics in the 1960s and 1970s and the rise of flexible accumulation and production. Flexibility in the marketplace removed the power of labor, and citizen activism moved to the community, where women always had an important stake.

4. S.H.A.R.E. (Self-Help and Resource Exchange) is a program designed to bring together low-income families and individuals in a community of involvement and responsibility; time can be donated in exchange for food, clothing, and other resources. It also serves as a source of information about other self-help programs.

Bibliography

Adams, Paul, Liela Berg, Nan Berger, Michael Duane, A. S. Neill, and Robert Ol-
    lendorff. 1971. *Children's Rights: Towards the Liberation of the Child.* New York:
    Praeger.
Aitken, Stuart C. 1994. *Putting Children in Their Place.* Washington, D.C.: Asso-
    ciation of American Geographers.
Aitken, Stuart C., and Thomas Herman. 1997. Gender, Power and Crib Geogra-
    phy: From Transitional Spaces to Potential Places. *Gender, Place and Culture* 4
    (1): 63–88.
Aitken, Stuart C., and Suzanne Michel. 1995 . Who Contrives the "Real" in GIS?
    Geographic Information, Planning and Critical Theory. *Cartography and Geo-
    graphic Information Systems* 22 (1): 17–29.
Aitken, Stuart C., and Leo E. Zonn. 1993. Weir(d) Sex: Representations of Gen-
    der-Environment Relations in Peter Weir's *Picnic at Hanging Rock* and *Gallipoli.*
    Society and Space 11: 191–212.
———, eds. 1994. *Place, Power, Situation and Spectacle: A Geography of Film.*
    Lanham, Md.: Rowman & Littlefield.
———. 1994. Representing the Place Pastiche. In *Place, Power, Situation and Spec-
    tacle: A Geography of Film,* ed. Stuart C. Aitken and Leo E. Zonn, 3–25.
    Lanham, Md.: Rowman & Littlefield.
Alanen, Leena. 1994. Gender and Generation: Feminism and the "Child Question."
    In *Childhood Matters: Social Theory, Practice and Politics,* ed. Jens Qvortrup,
    Marjatta Bardy, Giovanni Sgritta, and Helmut Wintersberger, 27–42. Aldershot,
    U.K.: Averbury Press.
Anderson, Benedict. 1991. *Imagined Communities.* London: Verso.
Appleton, Lynn. 1995. The Gender Regimes of American Cities. In *Gender in Ur-
    ban Research,* ed. Judith A. Garber and Robyne S. Turner, 44–59. Thousand
    Oaks, Calif.: Sage.
Ariès, Philippe. 1962. *Centuries of Childhood: A Social History of Family Life.* Trans-
    lated by R. Baldick. New York: Knopf.
Ayto, John. 1990. *Dictionary of Word Origins.* New York: Arcade.
Barthes, Roland. 1972. *Mythologies.* New York : Hill & Wang.
Baudrillard, Jean. 1989. *America.* London: Verso.
Bell, David, Jon Binnie, Julia Cream, and Gill Valentine. 1994. All Hyped Up and
    No Place to Go. *Gender, Place and Culture* 1 (1): 31–47.
Benhabib, Seyla. 1992. *Situating the Self: Gender, Community and Postmodernism in
    Contemporary Ethics.* New York: Routledge.
Berkner, Lutz K. 1972. The Stem Family and the Development Cycle of the Peasant

Household: An Eighteenth-Century Austrian Example. *American Historical Review* 77: 131–47.

———. 1975. The Use and Misuse of Census Data for the Historical Analysis of Family Structure. *Journal of Interdisciplinary History* 5: 721–38.

Berman, Marshall. 1982. *All That Is Solid Melts into Air: The Experience of Modernity.* New York: Simon & Schuster.

Bernard, Jessie. 1981. The Good Provider Role: Its Rise and Fall. *American Psychologist* 36: 1–12.

Bernardes, Jon. 1985a. Do We Really Know What 'The Family' Is? In *Family and Economy in Modern Society,* ed. P. Close and R. Collins, 192–211. London: Macmillan.

———. 1985b. "Family Ideology": Identification and Exploration. *Sociological Review* 33: 275–97.

———. 1993. Responsibilities in Studying Postmodern Families. *Journal of Family Issues* 14: 35–49.

Bielby, D. D., and W. T. Bielby. 1988. Women's and Men's Commitment to Paid Work and Family: Theories, Models and Hypotheses. In *Women and Work: An Annual Review,* ed. B. A. Gutek, A. H. Stromberg, and L. Larwood, 249–63. Beverly Hills, Calif.: Sage.

Blau, F., and M. Ferber. 1985. Women in the Labor Market: The Last Twenty Years. In *Women and Work: An Annual Review,* ed. L. Larwood, A. H. Stromberg, and B. A. Gutek, 19–49. Beverly Hills, Calif.: Sage.

Bloch, Ruth. 1978. American Feminine Ideals in Transition: The Rise of the Moral Mother, 1785–1815. *Feminist Studies* 2: 100–126.

Bondi, Liz. 1993. Locating Identity Politics. In *Place and the Politics of Identity,* ed. Michael Keith and Steve Pile, 84–101. London: Routledge.

———. 1997. In Whose Words? On Gender Identities, Knowledge and Writing Practices. *Transactions of the British Institute of Geographers.* Forthcoming.

Bowlby, John. 1951. *Maternal Care and Mental Health.* Geneva: World Health Organization.

Boys, J. 1984. Making Out: The Place of Women Outside the Home. In *Making Space: Women in the Man-Made Environment.* Matrix Architects: Book Group. London: Pluto Press.

Burgess, Ernest. 1926. The Family as a Unity of Interacting Personalities. *The Family* 7: 3–9.

———. 1973. *On Community, Family, and Delinquency.* Chicago: University of Chicago Press.

Butler, Judith. 1992. Contingent Foundations: Feminism and the Question of "Postmodernism." In *Feminists Theorize the Political,* ed. J. Butler and J. Scott, 3–21. New York: Routledge.

———. 1993. *Bodies That Matter: On the Discursive Limits of "Sex."* New York: Routledge.

Chaytor, Miranda. 1980. Household and Kinship: Ryton in the Late Sixteenth and Early Seventeenth Centuries. *History Workshop* 10.

Cheal, David. 1993. Unity and Difference in Postmodern Families. *Journal of Family Issues* 14: 5–19.

Chodorow, Nancy. 1974. Family Structure and Feminine Personality. In *Women, Culture and Society,* ed. M. Zimbalist, M. Rosaldo, and L. Lamphere, 106–27. Stanford, Calif.: Stanford University Press.

———. 1978. *The Reproduction of Mothering.* Berkeley: University of California Press.

———. 1989. *Feminism and Psychoanalytic Theory.* New Haven, Conn.: Yale University Press.

————. 1994. *Femininities, Masculinities, Sexualities: Freud and Beyond.* Lexington: University of Kentucky Press.

Chodorow, Nancy, and Susan Contratto. 1992. The Fantasy of the Perfect Mother. In *Rethinking the Family: Some Feminist Questions,* ed. Barrie Thorne and Marilyn Yalom, 191–214. Boston: Northeastern University Press.

Clifford, James. 1988. *The Predicament of Culture: Twentieth Century Ethnography, Literature and Art.* Cambridge, Mass.: Harvard University Press.

Collier, J., M. Z. Rosaldo, and S. Yanagisako. 1992. Is There a Family? New Anthropological Views. In *Rethinking the Family: Some Feminist Questions,* ed. Barrie Thorne and Marilyn Yalom, 31–59. Boston: Northeastern University Press.

Cowan, Susan, and Mary F. Katzenstein. 1988. The War over the Family Is Not over the Family. In *Feminism, Children and the New Families,* ed. Sanford M. Dornbusch and Myra H. Strober, 25–46. New York: Guilford Press.

Crang, Mike. 1992. The Politics of Polyphony: Re-configurations in Geographical Authority. *Society and Space* 10: 527–49.

Croghan, R. 1991. First-Time Mothers' Accounts of Inequality in the Division of Labour. *Feminism & Psychology* 1: 221–46.

Deem, R. 1986. *All Work and No Play: The Sociology of Women's Leisure.* Bristol, Pa.: Open University Press.

Del Castillo, Adelaida R. 1993. Covert Cultural Norms and Sex/Gender Meaning: A Mexico City Case. *Urban Anthropology* 22 (3–4): 237–58.

Demos, John. 1970. *A Little Commonwealth: Family Life in Plymouth Colony.* New York: Oxford University Press.

————. 1986. *Past, Present and Personal: The Family and the Life Course in American History.* New York: Oxford University Press.

Dennis, Richard, and Stephen Daniels. 1994. "Community" and the Social Geography of Victorian Cities. In *Time, Family and Community: Perspectives on Family and Community History,* ed. Michael Drake, 201–24. Oxford: Blackwell.

Donzelot, Jacques. 1979. *The Policing of Families.* Translated from the French by Robert Hurley. New York : Pantheon Books.

Durham, M. 1991. *Sex and Politics: The Family and Morality in the Thatcher Years.* Basingstoke, U.K.: Macmillan.

Durkheim, Émile. [1893] 1964. *The Division of Labor in Society.* Translated by G. Simpson. New York: Free Press.

Durkheim, Émile and Marcel Mauss. [1903] 1963. *Primitive Classifications.* Translated from the French and edited with an introduction by Rodney Needham. Chicago: University of Chicago Press.

Dyck, Isabel. 1990. Space, Time and Renegotiating Motherhood: An Exploration of the Domestic Workplace. *Society and Space* 8: 459–83.

Eggebeen, David, and Peter Uhlenberg. 1985. Changes in the Organization of Men's Lives: 1960–1980. *Family Relations* 34 (2): 251–57.

Ehrenreich, B. 1983. *The Hearts of Men: American Dreams and the Flight from Commitment.* Garden City, N.Y.: Anchor Books.

Elshtain, Jean B. 1990. The Family in Political Thought: Democratic Politics and the Question of Authority. In *Fashioning Family Theory,* ed. J. Sprey, 51–66. Newbury Park, Calif.: Sage.

Engels, Friedrich. [1845] 1968. *The Condition of the Working Class in England.* Translated and edited by W. O. Henderson and W. H. Chaloner. Stanford, Calif.: Stanford University Press.

England, Kim. 1993. Suburban Pink Collar Ghettos: The Spatial Entrapment of Women. *Annals of the Association of American Geographers* 83 (2): 225–42.

Epstein, Julia, and Kristina Straub, eds. 1991. *Body Guards: The Cultural Politics of Gender Ambiguity*. New York: Routledge.

Everhart, Robert. 1983. *Reading, Writing and Resistance: Adolescence and Labor in a Junior High School*. Boston: Routledge.

Fagnani, Jeanne. 1983. Women's Commuting Patterns in the Paris Region. *Tijdschrift woor Economische en Sociale Geografie* 74: 12–24.

———. 1993. Life Course and Space, Dual Careers and Residential Mobility among Upper Middle-Class Families in the Ill-de-France Region. In *Full Circles: Geographies of Women over the Life Course*, ed. Cindi Katz and Jan Monk, 171–87. New York: Routledge.

Fava, Sylvia. 1980. Women's Place in the New Suburbia. In *New Spaces for Women*, ed. Gerda R. Wekerle, Rebecca Peterson, and David Morley, 129–49. Boulder, Colo.: Westview Press.

Figueira-McDonough, Josefina, and Rosemary Sarri, eds. 1987. *The Trapped Women*. Newbury Park, Calif.: Sage.

Fitzsimmons, Margaret. 1989. The Matter of Nature. *Antipode* 21: 213–25.

Flax, Jane. 1978. The Conflict between Nurturance and Autonomy in Mother-Daughter Relationships and within Feminism. *Feminist Studies* 2: 171–89.

———. 1983. Contemporary American Families: Decline or Transformation? In *Families, Politics, and Public Policy: A Feminist Dialogue on Women and the State*, ed. Irene Diamond, 21–40. New York:

———. 1990. *Thinking Fragments: Psychoanalysis, Feminism, and Postmodernism in the Contemporary West*. Berkeley: University of California Press.

———. 1993. *Disputed Subjects: Essays on Psychoanalysis, Politics, and Philosophy*. New York and London: Routledge.

Foord, Jo, and Nicky Gregson. 1986. Patriarchy: Towards a Reconceptualization. *Antipode* 18: 181–211.

Foucault, Michel. 1977. *Discipline and Punishment*. New York: Penguin Books.

———. 1980. *Power/Knowledge*. New York: Pantheon Books.

Fraser, Elizabeth, and Nicola Lacey. 1993. *The Politics of Community: A Feminist Critique of the Liberal-Communitarian Debate*. Toronto: University of Toronto Press.

Frønes, Ivar. 1994. Dimensions of Childhood. In *Childhood Matters: Social Theory, Practice and Politics*, ed. Jens Qvortrup, Marjatta Bardy, Giovanni Sgritta, and Helmut Wintersberger, 145–64. Aldershot, U.K.: Avebury Press.

Furstenberg, Frank F. 1988. Good Dads–Bad Dads: Two Faces of Fatherhood. In *The Changing American Family and Public Policy*, ed. Andrew J. Cherlin, 193–218. Washington, D.C.: Urban Institute Press.

Gans, Herbert. 1962. *The Urban Villagers*. New York: Free Press.

———. 1967. *The Levittowners: Ways of Life and Politics in a New Suburban Community*. New York: Vintage Books.

Garber, Judith A. 1995. Defining Feminist Community: Place, Choice, and the Urban Politics of Difference. In *Gender in Urban Research*, ed. Judith A. Garber and Robyne S. Turner, 24–43. Thousand Oaks, Calif.: Sage.

Geertz, Clifford. 1983. *Local Knowledge*. New York: Basic Books.

Gieve, K. 1987. Rethinking Feminist Attitudes towards Motherhood. *Feminist Review* 25: 38–45.

Giddens, Anthony. 1990. *The Consequences of Modernity*. Stanford, Calif.: Stanford University Press.

Gilbert, Melissa R. 1994. The Politics of Location: Doing Feminist Research at "Home." *Professional Geographer* 46 (1): 90–96.

Gilligan, Carol. 1982. *In a Different Voice*. Cambridge, Mass.: Harvard University Press.

Glasser, Ralph. 1977. *The Net and the Quest*. London: Temple Smith.

———. 1986. *Growing up in the Gorbals*. London: Pan Books.

Glenn, Evelyn Nakano. 1987. Gender and the Family. In *Analyzing Gender: A Handbook of Social Science Research*, ed. Beth Hess and Myra Marx Feree, 348–80. Newbury Park, Calif.: Sage.

Goldscheider, Frances V., and Linda J. Waite. 1991. *New Families, No Families? The Transformation of the American Home*. A RAND Study. Berkeley: University of California Press.

Gordon, Linda. 1992. Why Nineteenth-Century Feminists Did Not Support "Birth Control" and Twentieth-Century Feminists Do: Feminism, Reproduction and the Family. In *Rethinking the Family: Some Feminist Questions*, ed. Barrie Thorne and Marilyn Yalom, 140–54. Boston: Northeastern University Press.

Gottlieb, Beatrice. 1993. *The Family in the Western World: From the Black Death to the Industrial Age*. New York: Oxford University Press.

Gregory, Derek. 1994. *Geographical Imaginations*. Cambridge, Mass.: Blackwell.

Gregory, Derek, and John Urry. 1985. Introduction. In *Social Relations and Spatial Structures*, ed. Derek Gregory and John Urry, 1–8. London: Macmillan.

Gregson, Nicky, and Michelle Lowe. 1993. Renegotiating the Domestic Division of Labor. *Sociological Review* 41(3): 475–505.

———. 1995. "Home"-making: On the Spatiality of Daily Social Reproduction in Contemporary Middle-Class Britain. *Transactions of the Institute of British Geographers* 20: 224–35.

Habermas, Jürgen. 1979. *Communication and the Evolution of Society*. Boston: Beacon Press.

———. 1984. *The Theory of Communicative Action*. Vol. 1. Boston: Beacon Press.

———. 1987a. *The Philosophical Discourse of Modernity: Twelve Lectures*. Cambridge, Mass.: MIT Press.

———. 1987b. *The Theory of Communicative Action*. Vol. 2. Boston: Beacon Press.

———. 1989. *The Structural Transformation of the Public Sphere*. Cambridge, Mass.: MIT Press.

Hanson, Susan, and Peter Hanson. 1980. Gender and Urban Activity Patterns in Uppsala, Sweden. *Geographical Review* 70: 291–99.

Hanson, Susan, and I. Johnston. 1985. Gender Difference in Worktrip Length. *Urban Geography* 6: 193–219.

Hanson, Susan, and Geraldine Pratt. 1988. Reconceptualizing the Links between Home and Work in Urban Geography. *Economic Geography* 64(4): 299–321.

———. 1994. Commentary on "Suburban Pink Collar Ghettos: The Spatial Entrapment of Women" by Kim England. *Annals of the Association of American Geographers* 84(3): 500–504.

———. 1995. *Gender, Work, and Space*. New York: Routledge.

Hareven, Tamara K. 1982. *Family Time and Industrial Time: The Relation between the Family and Work in a New England Industrial Community*. Cambridge, Mass.: Harvard University Press.

———. 1994. Recent Research on the History of the Family. In *Time, Family and Community: Perspectives on Family and Community History*, ed. Michael Drake, 13–43. Oxford: Blackwell.

Hart, Roger. 1984. The Geography of Children and Children's Geography. In *Environmental Perception and Behavior: An Inventory and Prospect*, ed. T. F. Saarinen, D. Seamon, and J. L. Sell, 99–129. Chicago: University of Chicago Press.

Harvey, David. 1989. *The Condition of Postmodernity*. Cambridge, Mass.: Blackwell.

————. 1993. From Space to Place and Back Again: Reflections on the Condition of Postmodernity. In *Mapping the Futures: Local Cultures, Global Change*, ed. Jon Bird, Bary Curtis, Tim Putnam, George Robertson, and Lisa Tickner, 3–29. London: Routledge.

Hayden, Dolores. 1984. *Redesigning the American Dream: The Future of Housing, Work, and Family Life*. New York: Norton.

Hayford, Alison M. 1974. The Geography of Women: An Historical Introduction. *Antipode* 6: 1–19.

Herman, Thomas. 1995. Exploratory Data Analysis of a Sample Population. Unpublished manuscript.

Hengst, Herbert. 1987. The Liquidation of Childhood—An Objective Tendency. *International Journal of Sociology* 17: 58–80.

Healey, Patsy. 1992. Planning through Debate: The Communicative Turn in Planning Theory. *Transportation Planning Research* 63(2): 143–62.

Hey, David. 1993. *The Oxford Guide to Family History*. Oxford: Oxford University Press.

Hochschild, Arlie and Anne Machung. 1989. *The Second Shift: Working Parents and the Revolution at Home*. New York: Viking Press.

Howard, Ebenezer. 1902. *Garden Cities of Tomorrow*, 3d ed. London: Faber and Faber. Originally published in 1898 as *To-morrow: A Peaceful Path to Real Reform*.

Ireland, Mardy S. 1993. *Re-conceiving Women: Separating Motherhood from Female Identity*. New York: Guilford Press.

Jablonsky, Thomas. 1993. *Pride in the Jungle: Community and Everyday Life in Back of the Yards Chicago*. Baltimore: Johns Hopkins University Press.

Jackle, John, and David A. Wilson. 1992. *Derelict Landscapes: The Wasting of America's Built Environment*. Lanham, Md.: Rowman & Littlefield.

Jacobs, Jane. 1961. *The Death and Life of Great American Cities: The Failure of Town Planning*. New York: Random House.

Jameson, Fredric. 1984. Postmodernism, or the Cultural Logic of Late Capitalism. *New Left Review* 146: 53–92.

————. 1992. *The Geopolitical Aesthetic: Cinema and Space in the World System*. Bloomington: Indiana University Press.

Jensen, An-Magritt. 1994. The Feminization of Childhood. In *Childhood Matters: Social Theory, Practice and Politics*, ed. Jens Qvortrup, Marjatta Bardy, Giovanni Sgritta, and Helmut Wintersberger, 59–75. Aldershot, U.K.: Averbury Press.

Johnston, Lynda, and Gill Valentine. 1995. Wherever I Lay My Girlfriend, That's My Home: The Performance and Surveillance of Lesbian Identities in Domestic Environments. In *Mapping Desire*, ed. David Bell and Gill Valentine, 99–113. London: Routledge.

Johnston-Anumonwo, Ibipo. 1992. The Influence of Household Type on Gender Differences in Work Trip Distance. *Professional Geographer* 44: 161–69.

Johnston-Anumonwo, Ibipo, Sara McLafferty, and Valerie Preston. 1995. Gender, Race and the Spatial Context of Women's Employment. In *Gender in Urban Research*, ed. Judith A. Garber and Robyne S. Turner, 236–55. Thousand Oaks, Calif.: Sage.

Kandiyotti, Deniz. 1988. Bargaining with Patriarchy. *Gender & Society* 2: 274–90.

Katz, Cindi. 1991a. Cable to Cross a Curse: The Everyday Practices of Resistance and Reproduction among Youth in New York City. Unpublished manuscript, Department of Environmental Psychology, City University of New York.

————. 1991b. Sow What You Know: The Struggle for Social Reproduction in Rural Sudan. *Annals of the Association of American Geographers* 81(3): 488–514.

————. 1992. All the World Is Staged: Intellectuals and the Projects of Ethnography. *Environment and Planning D: Society and Space* 10: 495–510.

————. 1993. Growing Girls/Closing Circles. In *Full Circles: Geographies of Women over the Lifecourse*, ed. Cindi Katz and J. Monk, 88–106. London: Routledge.

————. 1994. Playing the Field: Questions of Fieldwork in Geography. *Professional Geographer* 46 (1): 67–72.

Kennedy, David. 1991. The Young Child's Experience of Space and Child Care Center Design: A Practical Meditation. *Children's Environments Quarterly* 8 (1): 37–48.

Kincheloe, Joe L., and Peter L. McLaren. 1994. Rethinking Critical Theory and Qualitative Research. In *The Handbook of Qualitative Methods*, ed. Norman Denzin and Yvonna Lincoln, 138–57. Newbury Park, Calif.: Sage.

Kirby, Kathleen. 1996. *Indifferent Boundaries: Spatial Concepts of Human Subjectivity*. New York: Guildford Press.

Klodawsky, Fran, and Aron Spector. 1990. New Families, New Housing Needs, New Urban Environments: The Case of Single-Parent Families. In *Life Spaces: Gender, Household, Employment*, ed. Caroline Andrew and Beth Moore Milroy, 141–58. Vancouver: University of British Columbia Press.

Knox, Paul. 1991. The Restless Urban Landscape: Economic and Sociocultural Change in the Transformation of Metropolitan Washington, D.C. *Annals of the Association of American Geographers* 81 (2): 181–209.

————. 1994a. The Stealthy Tyranny of Community Spaces. *Environment and Planning A* 26 (2): 170–73.

————. 1994b. *Urbanization: An Introduction to Urban Geography*. Englewood Cliffs, N.J.: Prentice-Hall.

Kobayashi, Audrey. 1994. Coloring the Field: Gender, "Race," and the Politics of Fieldwork. *Professional Geographer* 46 (1): 73–80.

Lacan, Jacques. 1978. *The Four Fundamental Concepts of Psycho-analysis*. Translated by Alan Sheridan. New York: Norton.

Lacey, Nicola. 1994. Community, Identity and Power: Some Thoughts on Women and Law in Eastern Europe. Paper presented at the annual meetings of the Association of American Geographers, San Francisco.

Lamb, Michael E. 1981. Historical Perspectives on the Father's Role. In *The Role of the Father in Child Development*, 2d ed., ed. Michael E. Lamb, 1–18. New York: Wiley.

————, ed. 1987. *The Father's Role: Applied Perspectives*. New York: Wiley.

Laqueur, Thomas W. 1990. *Making Sex: Body and Gender from the Greeks to Freud*. Cambridge, Mass.: Harvard University Press.

————. 1992. The Facts of Fatherhood. In *Rethinking the Family: Some Feminist Questions*, ed. Barrie Thorne and Marilyn Yalom, 155–75. Boston: Northeastern University Press.

Lasch, Christopher. 1977. *Haven in a Heartless World: The Family Besieged*. New York: Basic Books.

Lash, Scott, and John Urry. 1994. *Economies of Signs and Space*. London: Sage.

Laslett, Peter. 1977. *Family Life and Illicit Love in Earlier Generations*. Cambridge: Cambridge University Press.

Laslett, Peter, and Richard Wall, eds. 1972. *Household and Family in Past Time*. Cambridge: Cambridge University Press.

Lee, Terrence. 1968. Urban Neighborhood as a Socio-spatial Schemata. *Human Relations* 21: 246–68.

Lefebvre, Henri. 1984. *Everyday Life in the Modern World*. New Brunswick, N.J.: Transaction.

———. 1991. *The Production of Space.* Translated by Donald Nicholson-Smith. Oxford: Blackwell.

Lewis, Robert A., and Marvin B. Sussman. 1986. *Men's Changing Roles in the Family.* New York: Haworth Press.

Lewis, S.N.C., and C. L. Cooper. 1988. Stress in Dual-Earner Families. In *Women and Work: An Annual Review,* ed. B. A. Gutek, A. H. Stromberg, and L. Larwood, 139–68. Beverly Hills, Calif.: Sage.

Loscocco, Karyn, and Joyce Robinson. 1991. Barriers to Women's Small-Business Success in the U.S. *Gender and Society* 5: 511–32.

McCann, Eugene. 1995. Neotraditional Developments: The Anatomy of a New Urban Form. *Urban Geography* 16 (3): 210–233.

Maccoby, Eleanor E. 1988. Gender as a Social Category. *Developmental Psychology* 24: 755–65.

———. 1990. Gender and Relationships: A Developmental Account. *American Psychologist* 45: 513–20.

Maccoby, Eleanor E., and J. A. Martin. 1983. Socialization in the Context of the Family: Parent-Child Interaction. In *Handbook of Child Psychology,* ed. E. Mavis Hetherington, 1–102. San Francisco: Jossey-Bass.

McDowell, Linda. 1983. Towards an Understanding of the Gender Division of Urban Space. *Environment and Planning D: Society and Space* 1: 59–72.

———. The Baby and the Bathwater: Diversity, Deconstruction and Feminist Theory in Geography. *Geoforum* 22 (2): 123–33.

McKenzie, Evan. 1994. *Privatopia: Homeowners Associations and the Rise of Residential Private Government.* New Haven, Conn.: Yale University Press.

Mackenzie, Suzanne. 1988. Building Women, Building Cities. In *Life Spaces: Gender, Household, Employment,* ed. Caroline Andrew and Beth Moore Milroy, 13–30. Vancouver: University of British Columbia Press.

———. 1989. Restructuring the Relations of Work and Life: Women as Environmental Actors, Feminism as Geographic Analysis. In *Remaking Human Geography,* ed. A. Kobayashi and Suzanne Mackenzie, 40–61. Boston: Unwin Hyman.

Mackenzie, Suzanne, and Damaris Rose. 1983. Industrial Change, the Domestic Economy and Home Life. In *Redundant Spaces in Cities and Regions: Studies in Industrial Decline and Social Change,* ed. J. Anderson, S. Duncan, and R. Hudson, 155–200. London: Academic Press.

McLafferty, Sally, and Valerie Preston. 1991. Gender, Race and Commuting among Service Sector Workers. *Professional Geographer* 43 (1): 1–15.

Madden, J. 1981. Why Women Work Closer to Home. *Urban Studies* 18: 181–94.

Malinkowski, B. 1913. *The Family among the Australian Aborigines.* London: University of London Press.

Marcus, George E., and Martin M. J. Fischer. 1986. *Anthropology as Cultural Critique: An Experimental Moment in the Human Sciences.* Chicago: University of Chicago Press.

Marsh, Margaret. 1990. *Suburban Lives.* New Brunswick, N.J.: Rutgers University Press.

Marston, Sallie A. 1995. "Female Citizens": Middle Class Women and the Domestic Management Movement in 19th Century Urban America. Invited lecture at the University of Colorado, Boulder.

Marston, Sallie A., and M. Saint-Germaine. 1991. Urban Restructuring and the Emergence of New Political Groupings: Women and Neighborhood Activism in Tucson, Arizona, U.S.A. *Geoforum* 22(2): 122–34.

Massey, Doreen. 1994. *Space, Place and Gender.* Minneapolis: University of Minnesota Press.

————. 1995. Masculinity, Dualisms and High Technology. *Transactions of the Institute of British Geographers* 20 (4): 487–99.

Michelson, William H. 1986. Basic Dimensions for the Analysis of Behavioral Potential in the Urban Environment II: An Update on Methodological and Substantive Results. In *Cross Cultural Research in Environment and Behavior*, Proceedings of the 2d U.S.-Japan Seminar, ed. W. H. Ittelson, M. Asai, and M. Ker, 195–207. Tucson: Department of Psychology, University of Arizona.

————. 1988. Divergent Convergence: The Daily Routines of Employed Spouses as a Public Affairs Agenda. In *Life Spaces: Gender, Household, Employment*, ed. Caroline Andrew and Beth Moore Milroy, 81–101. Vancouver, University of British Columbia Press.

Miller, Byron. 1992. Collective Action and Rational Choice: Place, Community, and the Limits to Individual Self-Interest. *Economic Geography* 69 (1): 22–62.

Miller, Jeanne, and Howard Garrison. 1982. Sex Roles: The Division of Labor at Home and in the Workplace. *American Sociological Review* 8: 237–62.

Miller, Zane L. 1981. The Role and Concept of Neighborhood in American Cities. In *Community Organization for Urban Social Change: A Historical Perspective*, ed. Robert Fisher and Peter Romanosky, 3–30. Westport, Conn.: Greenwood Press.

Moore, Susanne. 1988. Getting a Bit of the Other—The Pimps of Postmodernism. In *Postmodernism and Its Discontents: Theories and Practices*, ed. E. A. Kaplan, 88–104. London: Verso.

Monk, Janice. 1992. Gender in the Landscape: Expressions of Power and Meaning. In *Inventing Places: Studies in Cultural Geography*, ed. Kay Anderson and Fay Gale, 123–38. White Plains, N.Y.: Longman Cheshire.

Morgan, S. Philip, and Linda J. Waite. 1987. Parenthood and the Attitudes of Young Adults. *American Sociological Review* 52 (4): 541–47.

Mouffe, Chantal. 1991. Democratic Citizenship and the Political Community. In *Community at Loose Ends*, ed. Miami Theory Collective, 369–84. Minneapolis: University of Minnesota Press.

————. 1992. Feminism, Citizenship and Radical Democratic Practices. In *Feminists Theorize the Political*, ed. Judith Butler and J. W. Scott, 369–84. New York: Routledge.

Mumford, Lewis. 1961. *The City in History: Its Origins, Its Transformations, and Its Prospects*. New York: Harcourt Brace Jovanovich.

Murdock, George Peter. 1949. *Social Structure*. New York: Free Press.

Nast, Heidi. 1994. Women in the Field: Critical Feminist Methodologies and Theoretical Perspectives. *Professional Geographer* 46(1): 54–66.

Nelson, Karen. 1986. Female Labor Supply Characteristics and the Suburbanization of Low-Wage Office Work. In *Production, Work and Territory*, ed. Allen Scott and Mike Storper, 169–71. London: Allen & Unwin.

Norwood, Vera, and Janice Monk. 1987. *The Desert Is Nobody*. New Haven, Conn.: Yale University Press.

Okin, Susan Moller. 1989. *Justice, Gender and the Family*. New York: Basic Books.

Oldman, David. 1994. Adult-Child Relations as Class Relations. In *Childhood Matters: Social Theory, Practice and Politics*, ed. Jens Qvortrup, Marjatta Bardy, Giovanni Sgritta, and Helmut Wintersberger, 43–58. Aldershot, U.K.: Averbury Press.

Park, Robert, and Ernest Burgess. 1967. *The City*. Chicago: University of Chicago Press.

Parsons, Talcott. 1955. The American Family: Its Relation to Personality and to the Social Structure. In *Family, Socialization, and Interaction Process*, ed. Talcott Parsons and R. F. Bales, 3–34. New York: Free Press.

————. 1971. The Normal American Family. In *Readings on the Sociology of the Family*, ed. B. N. Adams and T. Weirath, 53–66. Chicago: Markham.

Perry, Clarence. 1939. *Housing for the Machine Age*. New York: Russell Sage Foundation.

Pickles, John. 1986. Crazy Dualisms and Conceptual Leakage; or How We Can Continue to Misunderstand Human Geography. *Ohio Geographers* 14: 75–84.

Pleck, Joseph H. 1985. *Working Wives/Working Husbands*. Beverly Hills, Calif.: Sage.

Popenoe, David. 1988. *Disturbing the Nest: Family Change and Decline in Modern Societies*. New York: Aldine de Gruyter.

————. 1993. American Family Decline, 1960–1990: A Review and Appraisal. *Journal of Marriage and Family* 55: 527–55.

Poster, Mark. 1978. *Critical Theory of the Family*. New York: Seabury Press.

————. 1993. *Politics, Theory and Contemporary Culture*. New York: Columbia University Press.

Postman, David. 1982. *The Disappearance of Childhood*. New York: Delacorte Press.

Pred, Allan. 1986. *Place, Practice and Structure: Social and Spatial Transformations in S. Sweden 1750–1850*. Cambridge: Polity Press.

Presser, Harriet. 1988. Shift Work and Child Care among Young Dual-Earner American Parents. *Journal of Marriage and the Family* 50: 133–48.

Rich, Adrienne. 1977. *Of Woman Born: Motherhood as Experience and Institution*. New York: Norton.

Roan, S. 1991. Make Room for Dad. *Los Angeles Times*, November 3, E-1.

Roberts, Marion. 1990. Gender and Housing: The Impact of Design. *Built Environment* 16 (4): 257–68.

Roberts, Robert. 1971. *The Classic Slum*. Manchester: Manchester University Press.

————. 1976. *A Ragged Schooling*. Manchester: Manchester University Press.

Robertson, A. F. 1991. *Beyond the Family: The Social Organization of Human Reproduction*. Berkeley: University of California Press.

Robins, Kevin. 1996. *Into the Image: Culture and Politics in the Field of Vision*. London: Routledge.

Rose, Gillian. 1993. *Feminism and Geography*. Minnesota: University of Minnesota Press.

————. 1995. Making Space for the Female Subject of Feminism. In *Mapping the Subject: Geographies of Cultural Transformation*, ed. Steve Pile and Nigel Thrift, 332–54. London: Routledge.

Rossi, Alice. 1973. Maternalism, Sexuality and the New Feminisms. In *Contemporary Sexual Behavior*, ed. Joseph Zubin and John Money, 145–74. Baltimore: Johns Hopkins University Press.

Rouner, Leroy S. 1991. Introduction. In *On Community*, ed. Leroy S. Rouner, 1–11. Notre Dame, Ind.: University of Notre Dame Press.

Rousenau, Pauline. 1992. *Postmodern Theory and the Social Sciences*. Princeton, N.J.: Princeton University Press.

Ruddick, Sara. 1992. Thinking about Fathers. In *Rethinking the Family: Some Feminist Questions*, ed. Barrie Thorne and Marilyn Yalom, 176–90. Boston: Northeastern University Press.

Ruddick, Susan. 1996. *Young and Homeless in Hollywood*. New York: Routledge.

Rutherford, B., and G. Wekerle. 1988. Captive Rider, Captive Labor: Spatial Constraints on Women's Employment. *Urban Geography* 9: 116–37.

Sandel, Michael. 1982. *Liberalism and the Limits of Justice*. Cambridge: Cambridge University Press.

Sayer, Andrew. 1984. *Method in Social Science: A Realist Approach*. London: Routledge.

———. 1991. Behind the Locality Debate: Deconstructing Geography's Dualisms. *Environment and Planning A* 23: 283–308.

Seidler, Victor J. 1995. Men, Heterosexuality, and Emotional Life. In *Mapping the Subject: Geographies of Cultural Transformation*, ed. Steve Pile and Nigel Thrift, 170–191. London: Routledge.

Selznick, P. 1992. *The Moral Commonwealth: Social Theory and the Promise of Community*. Berkeley: University of California Press.

Shorter, Edward. 1977. *The Making of the Modern Family*. New York: Basic Books.

Sibley, David. 1995. Families and Domestic Routines: Constructing the Boundaries of Childhood. In *Mapping the Subject: Geographies of Cultural Transformation*, ed. Steve Pile and Nigel Thrift, 123–37. London: Routledge.

Silver, Catherine. 1982. *Frédéric Le Play: On Family, Work and Social Change*. Chicago: University of Chicago Press.

Simmel, Georg. [1902] 1950. The Metropolis and Mental Life. In *The Sociology of Georg Simmel*, translated by K. H. Wolff, 635–46. New York: Free Press.

Sjoberg, Gideon. 1965. Community. In *Dictionary of Sociology*, ed. J. Gold and W. L. Kolb. London: Tavistock.

Smith, C. A. and C. J. Smith. 1978. Locating Natural Neighbors in the Urban Community. *Area* 10: 102–10.

Smith, Dorothy E. 1993. The Standard North American Family Unit: SNAF as an Ideological Code. *Journal of Family Issues* 14: 50–65.

Smith, Michael. 1984. Becoming a Good and Competent Community. In *Rebuilding America's Cities: Roads to Recovery*, ed. Paul Porter and David Sweet, 123–42. New Brunswick, N.J.: Rutgers University Press.

Smith, Neil. 1992. Geography, Difference and the Politics of Scale. In *Postmodernism and the Social Sciences*, ed. Joe Doherty, Elspeth Graham, and Mo Malek, 57–79. London: Macmillan.

———. 1993. Homeless/Global: Scaling Places. In *Mapping the Futures: Local Cultures, Global Change*, ed. Jon Bird, Bary Curtis, Tim Putnam, George Robertson, and Lisa Tickner, 87–119. London: Routledge.

Smith, Neil, and Cindi Katz. 1993. Grounding Metaphor: Towards a Spatialized Politics. In *Place and the Politics of Identity*, ed. Michael Keith and Steve Pile, 67–83. London: Routledge.

Smout, T. C. 1969. *The History of the Scottish People 1560–1830*. Glasgow: William Collins.

Soja, Edward W. 1985. The Spatiality of Social Life: Towards a Transformative Retheorization. In *Social Relations and Spatial Structures*, ed. Derek Gregory and John Urry, 90–127. London: Macmillan.

———. 1989. *Postmodern Geographies: The Reassertion of Space in Critical Social Theory*. London: Verso.

———. 1996. *Thirdspace: Journeys to Los Angeles and Other Real-and-Imagined Places*. Cambridge, Mass.: Blackwell.

Stacey, Judith. 1990. *Brave New Families: Stories of Domestic Upheaval in Late Twentieth Century America*. New York: Basic Books.

———. 1992. Backwards toward the Postmodern Family: Reflections on Gender, Kinship and Class in the Silicon Valley. In *Rethinking the Family: Some Feminist Questions*, ed. Barrie Thorne and Marilyn Yalom, 91–117. Boston: Northeastern University Press.

———. 1993. Good Riddance to "The Family": A Response to David Popenoe. *Journal of Marriage and the Family* 55: 545–47.

Stacey, Margaret. 1960. *Tradition and Change: A Study of Banbury.* Oxford: Oxford University Press.

———. 1969. The Myth of Community Studies. *British Journal of Sociology* 20: 134–47.

Stanton, Barbara. 1986. The Incidence of Home Grounds and Experiential Networks. *Environment and Behavior* 18 (3): 299–329.

Swerdlow, Amy, Renate Bridenthal, Joan Kelly, and Phyllis Vine. 1989. *Families in Flux.* New York: Feminist Press.

Taylor, Charles. 1989. *Sources of the Self* Cambridge: Cambridge University Press.

Thorne, Barrie. 1987. Re-visioning Women and Social Change: Where Are the Children? *Gender and Society* 1: 85–109.

———. 1992. Feminism and the Family: Two Decades of Thought. In *Rethinking the Family: Some Feminist Questions,* ed. Barrie Thorne and Marilyn Yalom, 3–30. Boston: Northeastern University Press.

Till, Karen. 1993. Neotraditional Towns and Urban Villages: The Cultural Production of a Geography of "Otherness." *Environment and Planning D: Society and Space* 11: 709–32.

Tivers, Jaqueline. 1985. *Attached Women: The Daily Lives of Women with Young Children.* London and Sydney: Croom Helm.

———. 1988. Women with Young Children: Constraints on Activities in the Urban Environment. In *Women in Cities: Gender and the Urban Environment,* ed. J. Little, L. Peake, and P. Richardson. London: Macmillan.

Todd, Emmanuel. 1985. *The Explanation of Ideology: Family Structures and Social Systems.* Oxford: Blackwell.

———. 1987. *The Cause of Progress: Culture, Authority and Change.* Oxford: Blackwell.

Tönnies, Ferdinand. [1887] 1957. *Community and Society.* Translated by C. Loomis. East Lansing: Michigan State University Press.

U.S. Bureau of the Census. 1990. *1990 Census of Population and Housing.* Washington, D.C.: U.S. Department of Commerce.

Valentine, Gill. 1990. Women's Fear and the Design of Public Space. *Built Environment* 16 (4): 288–303.

———. 1993. Negotiating and Managing Multiple Sexual Identities: Lesbian Time-Space Geographies. *Transactions of the Institute of British Geographers* 18: 237–48.

———. 1996. "Oh Yes I Can." "Oh No You Can't:" Children and Parents' Understandings of Kids Competence to Negotiate Public Space Safely. Paper presented at the annual meetings of the Association of American Geographers. Charlotte, N.C.

Wachtel, Paul. 1983. *The Poverty of Affluence: A Psychological Portrait of the American Way of Life.* New York: Free Press.

Watson, Sophie. 1991. The Restructuring of Work and Home: Productive and Reproductive Relations. In *Housing and Labour Markets: Building the Connections,* ed. John Allen and Chris Hamnett, 136–154. London: Unwin Hyman.

Webber, Melvin. 1967. Order in Diversity: Places without Propinquity. In *Cities and. Space: The Future Use of Urban Land,* ed. Lowdon Wingo, Jr., 23–54. Baltimore: Johns Hopkins University Press.

Weber, Max. [1904–5] 1976. *The Protestant Ethic and the Spirit of Capitalism.* First published as a two-part article in *Archiv für Sozialwissenschaft und Sozialpolitik.* Translated by Talcott Parsons. Guildford, Surrey: George Allen & Unwin.

Wharton, Carol S. 1994. Finding Time for the "Second Shift": The Impact of Flexibility on Women's Double Days. *Gender & Society* 8: 189–205.

White, M. J. 1986. Sex Differences in Urban Commuting Behavior. *American Economic Review, Papers and Procedures* 76: 368–77.

———. 1987. *American Neighborhoods and Residential Differentiation.* New York: Russell Sage Foundation.

Williams, Raymond. 1985. *Keywords: A Vocabulary of Culture and Society.* New York: Oxford University Press.

Willis, Paul. 1981. *Learning to Labour: How Working Class Kids Get Working Class Jobs.* New York: Columbia University Press.

Wilson, Elizabeth. 1991. *The Sphinx and the City: Urban Life, the Control of Disorder, and Women.* Berkeley: University of California Press.

Winnicott, D. W. 1964. *The Child, the Family, and the Outside World.* Middlesex, U.K : Penguin Books.

———. 1965. *The Family and Individual Development.* New York: Basic Books.

———. 1971. *Playing and Reality.* London: Tavistock.

Wirth, Louis. 1938. Urbanism as a Way of Life. *American Journal of Sociology* 44: 1–24.

Wolfe, Maxime, and Leanne G. Rivlin. 1987. The Institutions in Children's Lives. In *Spaces for Children: The Built Environment and Child Development,* ed. C. S. Weinstein and T. G. David, 89–116. New York: Plenum.

Woodsworth, James S. [1911] 1972. *My Neighbor: The Study of City Conditions.* Toronto: University of Toronto Press.

Wrightson, Keith. 1982. *English Society 1580–1680.* London: Hutchinson.

Young, Iris Marion. 1990a. The Ideal of Community and the Politics of Difference. In *Feminism/Postmodernism,* ed. Linda Nicholson, 300–23. New York: Routledge.

———. 1990b. *Justice and the Politics of Difference.* Princeton, N.J.: Princeton University Press.

Young, Michael, and Peter Willmott. 1957. *Family and Kinship in East London.* Glencoe, Ill.: Free Press.

Zonn, Leo E., and Stuart C. Aitken. 1994. Of Pelicans and Men: Symbolic Landscapes, Gender and Australia's *Storm Boy.* In *Place, Power, Situation and Spectacle: A Geography of Film,* ed. Stuart C. Aitken and Leo E. Zonn, 137–159. Lanham, Md.: Rowman & Littlefield.

Zorbaugh, Harvey. 1929. *Gold Coast and the Slum.* Chicago: University of Chicago Press.

Index

About the Author

Stuart C. Aitken is professor of geography at San Diego State University in San Diego, California. His B.Sc., M.A., and Ph.D. were earned, respectively, at the University of Glasgow (Scotland), Miami University (Ohio), and the University of Western Ontario (Canada). His publications include *Putting Children in Their Place* (1994) and *Place, Power, Situation and Spectacle: A Geography of Film* (1994, coedited with Leo Zonn).